ANCIENT SEAS
OF SOUTHERN FLORIDA

ANCIENT SEAS OF SOUTHERN FLORIDA

THE GEOLOGY AND PALEONTOLOGY OF THE EVERGLADES REGION

Second Edition

Edward J. Petuch

and

David P. Berschauer

CRC Press
Taylor & Francis Group
Boca Raton London New York

CRC Press is an imprint of the
Taylor & Francis Group, an **informa** business

First edition published 2022
by CRC Press
6000 Broken Sound Parkway NW, Suite 300, Boca Raton, FL 33487-2742

and by CRC Press
2 Park Square, Milton Park, Abingdon, Oxon, OX14 4RN

Library of Congress Cataloging-in-Publication Data

Names: Petuch, Edward J., author. | Berschauer, David P., author.
Title: Ancient seas of southern Florida : the geology and paleontology of the Everglades region / Edward J. Petuch, David P. Berschauer.
Other titles: Geology of the Everglades and adjacent areas | Geology and paleontology of the Everglades region
Description: Second edition. | Boca Raton : CRC Press, 2021. | Revised edition of: The geology of the Everglades and adjacent areas / Edward J. Petuch, Charles E. Roberts. c2007. | Includes bibliographical references and index. | Summary: "Painting a complete picture of the history of the Everglades, this book provides an overview of the geology, paleontology, and paleoceanography of the region, covering almost 750 stratigraphic index fossils found in these beds"-- Provided by publisher.
Identifiers: LCCN 2021026739 (print) | LCCN 2021026740 (ebook) | ISBN 9780367566333 (hardcover) | ISBN 9781032072173 (paperback) | ISBN 9781003205944 (ebook)
Subjects: LCSH: Geology--Florida--Everglades Region. | Geomorphology--Florida--Everglades Region. | Paleontology--Florida--Everglades Region.
Classification: LCC QE100.E9 P48 2021 (print) | LCC QE100.E9 (ebook) | DDC 557.59/39--dc23
LC record available at https://lccn.loc.gov/2021026739
LC ebook record available at https://lccn.loc.gov/2021026740

ISBN: 9780367566333 (hbk)
ISBN: 9781032072173 (pbk)
ISBN: 9781003205944 (ebk)

Typeset in Times New Roman
by KnowledgeWorks Global Ltd.

Dedication

This book is dedicated to the following:

Linda J. Petuch, Eric and Rasa Petuch,

Brian Petuch and Kendra Berentsen, and Jennifer Petuch

and

Felicia Weisbrot Berschauer, Morgan Taylor Coker, and Jeremy Coker

ANCIENT SEAS OF SOUTHERN FLORIDA
The Geology and Paleontology of the Everglades Region

TABLE OF CONTENTS

Foreword to the First Edition

The Florida Platform, and its associated subsurface geology and surface landforms, is clearly a child of the sea. The geologic formations from the latter part of the Mesozoic through the Cenozoic were laid down in shallow seas and periodically reworked by marine, near-shore, and fluvial erosion/ deposition processes numerous times. The subsurface lithologies are complex and difficult to correlate due to limited data availability and lithologic similarity. Information is compiled in the form of water well cuttings, stratigraphic core tests, geophysical wire line logs, seismic reflection data, groundwater geochemistry, and limited surface exposures. Further, it is painstakingly difficult to understand and decipher the environmental changes that occurred throughout the geologic past, as the platform grew and was built by the various sedimentary processes.

The authors of *The Geology of the Everglades and Adjacent Areas* have done an outstanding job of compiling decades of data collected by their own field reconnaissance and other geoscientists. They have described the Cenozoic rock sequence units, both formally and informally, to establish the overall lithologic character, paleontology, age, and map extent of the various units. Their sequential paleoenvironmental interpretations and graphical figure representations are a significant contribution to the chronologic understanding of the development of the Florida Platform. They are to be congratulated for the innovative use of modern satellite technology and digital graphics to present their sequential paleogeography interpretations. This represents a significant contribution to the understanding of the development of the Florida carbonate platform, and it will assist other disciplines as they strive for better understanding of our groundwater resources, aquifer characterizations, paleoenvironmental interpretations, and historical/educational geology programs.

Walt Schmidt, Ph.D., PG #1
Florida State Geologist & Chief
Florida Geological Survey
(Florida State Geologist 1987-2008)

ACKNOWLEDGMENTS

Without the help and interest of the following individuals and organizations, this book could never have been produced. For allowing the senior author to conduct research on their properties, we wish to thank: **Enrique Tomeu** (Palm Beach Aggregates, Loxahatchee); **Edwin Rucks, Jr.** (Rucks Crystal Mine, Fort Drum); **Jennifer Williamson**, the **Dickerson Family**, **M. Todd Lybrand**, and **Jimmy Heimer** (Dickerson Aggregates Mines, Fort Drum and Indrio); **Andres Fanjul**, the **Fanjul Family**, **Mike Lorenz**, and **Alfonzo Alvarez** (Okeelanta Division, Florida Crystals, Inc., Everglades Agricultural Area); **Jose Dimas**, **Robert Roach**, **Richard Rhodes**, and **Iris Schellhorn** (Florida Rock Industries Naples Quarry, East Naples); **Ronald Capeletti** and the **Capeletti Family** (Capeletti Brothers Quarries, Hialeah); the **Richardson Family** and the **Ashland Asphalt and Petroleum Company** (APAC Pit, Sarasota); **Hugh Cannon** (Quality Aggregates Pit, Sarasota); **D.L. Brantley** (Brantley Pit, Arcadia); **Robin Weeks** (Weeks Pit, Lakeport); **Fred Smith** (Smith Pit, Okeechobee); **Howard A. Griffin** (Griffin Brothers Quarries, Fort Lauderdale); **Allen Ridgdill** (Ridgdill DOT Pit, Moore Haven); the **Bergeron Family** (Bergeron Star Ranch Pit, South Bay); the **Mecca Family** (Mecca Farms Pit, North Palm Beach); the **Bonita Grande Aggregates Pit** (Bonita Springs); **Five Stones Quarry** (Port Mayaca). Special thanks goes to **A. Kenneth ("Kenny") Brown**, Chokoloskee Island, for the generous loan of a boat, accommodations, and support for our research on the worm shell reefs of the Ten Thousand Islands, and to **Capt. Craig Daniels**, Chokoloskee Island, for guiding us through the southern Ten Thousand Islands and their worm shell reefs.

For the donation of valuable research specimens (many illustrated in this book) and for assistance with field work, we thank the following: **Dr. Anton Oleinik, Eddie Matchett, Clifford Swearingen, Maurice Guy, Melanie Briskin, Phyllis Diegel, Richard Duerr, Judy and Larry Haley, Patricia and Dale Stream, Violet and Charles Hertweck, Meta Jones, Dr. Charles Roberts, Lannie Gustafson, Jason Gabriel, Donald Asher, Brian Schnirel, Edward Volek, George Saloney, Gary Charney, Fonda and Herbert Waldron, Michael Bruggeman, Gary Leonard, Harry Yingst, Eric Kendrew, Robert Owens, Jimmy Kramp, Marilyn Barkley, Harry Hyaduck, Jonathan Arline, Joanna Arline, James Houbrick, Joseph Bucheck II, Richard Anderson, Vladimir Eismont, Susan Khan, Mary Mansfield, Diane and Richard Pennington, Cathy and William Dowling, Anthony Cinelli, William Aley, Rudolph Pascucci, Josiah Strauss, Jack Spengler, Stephen Bass, Freddy St. Jude, Theodore Davis, Walter ("Buddy") Ward, Mardie Drolshagen, Dr. Gregory Herbert, Dr. Gregory Dietl, Dr. John Whicher, Patrick McCue, and Dylan Sarmiento.**

For allowing us to illustrate some of their excellent photographs of exposures and sections, we thank **Dr. Anton Oleinik** (Florida Atlantic University), **Michael Bruggeman**, **Brian Schnirel**, and **Rudy Wobito**. Special thanks also go to **Robert F. Myers**, for taking the photographs of the fossil cowries and busycon whelks shown throughout this book, and to **Dr. Charles Roberts**, for creating the simulated satellite images of the Okeechobean Subseas.

View of the southern Everglades near Ochopee, Collier County, Florida, showing wet prairies composed of saw grass and *Scirpus* rushes, a small island covered with Sabal Palms (*Sabal palmetto*), and cypress forests along the horizon.

INTRODUCTION. The Everglades Region and the Okeechobean Sea

The principal physiographic feature of southern Florida, the Everglades, extends from north of Lake Okeechobee southward to the Gulf of Mexico. Within its central region, in Palm Beach, Hendry, Broward, Dade, and Collier Counties, the Everglades contains a subtropical wetland with the largest sawgrass prairies found anywhere in the United States. In reference to this seemingly-endless vista, the native Seminole people called this region the "Pa-Hay-Okee", or the "River of Grass". Myriads of small uninhabited tree islands are also scattered across the Everglades wet prairies, with each harboring a distinctive ecosystem that is composed of various combinations of cypresses, tropical hardwoods, and Sabal Palmettos. Everglades-type environments, including extensive cypress forests, also range into the surrounding counties of southern Florida, from Sarasota and Okeechobee Counties in the north to Lee and Monroe Counties in the south, making them the dominant ecological features across the southern part of the state. As described by Marjory Stoneman Douglas in her book *The Everglades: River of Grass* (1947), this special North American natural treasure has always fascinated, and most often repelled, non-native visitors and settlers. Even today, the word "Everglades" conjures up visions of steaming swamps filled with voracious alligators, swarms of disease-bearing mosquitoes, and venomous snakes. For thousands of years, only the Native Americans found refuge in these marshlands, adapting to the harsh conditions and creating their own unique lifestyles.

The Everglades and southern Florida that we see today are actually the latest iteration of a large sequence of marine environments that stretches back for tens of millions of years. At first glance, no one would consider the modern Everglades to be part of the marine realm; but shallow excavations across the region have exposed ancient marine worlds that are hidden just below the surface. These prehistoric sea floors, one stacked upon the other for thousands of meters, are geographically important, as they have created the topography that governs the water flow and sedimentary deposition within the modern Everglades. The spectacular subtropical world that was so poetically described by Marjory Stoneman Douglas is now known to be only a thin film of terrestrial sediments that covers the record of Florida's ancient seas. Any geologic or stratigraphic study of the Everglades region, consequently, must be approached from the oceanographic perspective. This oceanic connection becomes strikingly apparent when studying exposures of the amazingly rich Neogene marine fossil beds that are buried just beneath the Everglades region. In this book, we illustrate almost 800 of the most important and diagnostic stratigraphic index fossils found in these beds, including over 50 species of corals and almost 700 species of mollusks, along with echinoderms, crustaceans, echinoids, petrified wood, and aquatic vertebrates. The following book is, in essence, a second edition of *The Geology of the Everglades and Adjacent Areas* (Petuch & Roberts, 2007) and includes new, updated geological data and concepts, and an expanded iconography of stratigraphic index fossils. Based on the data gleaned from these fossils, we also offer a series of geomorphological visualizations, showing the possible appearances of the Florida Peninsula during the times when it was covered by tropical seas, from the Oligocene to the late Pleistocene.

Looking across the Everglades and the southern Florida of today, it is intriguing to imagine what the area may have looked like when it was completely covered by shallow marine environments. This ancient marine world, referred to here as the **Okeechobean Sea** (named for present-day Lake Okeechobee; see Petuch, 2004), was essentially land-locked for much of its later history, being separated from the Gulf of Mexico and Atlantic Ocean by a narrow semicircular chain of shallow oyster banks, coral reef tracts, and small coral islands. Imagine a cross between present-day Florida Bay, with its mangrove jungles and sea grass beds, and the Bahamas, with their coral cays and coral reef complexes; this would have typified the appearance of the Okeechobean Sea over the past 12 million years. As sea levels rose and fell over the millenia, the Okeechobean Sea area was completely under the influence of this large-scale rhythmic pattern, which is recorded in the geology as alternating episodes of terrestrial and marine sedimentation. For cyclostratigraphers and climatologists, these well-preserved marine sediments and fossil beds contain one of the most detailed records of Neogene global climate change and sea level fluctuations found anywhere on Earth.

Considering the serene beauty of the present-day Everglades area, it is hard to imagine that the predecessor Okeechobean Sea had been formed in the unimaginable chaos and destruction that resulted from an asteroid impact (astrobleme). At the end of the Eocene Epoch, around 35 million years ago, the southern end of what is now Chesapeake Bay in Virginia was struck by an asteroid with a diameter of approximately 13 km (8 miles), producing immense incandescent debris clouds and multiple mega-tsunami waves of over 500 m (1640 feet). These monster waves, which were generated over and over again for weeks on end due to the collapsing crater, roared down the coast and essentially obliterated every coastal geologic feature from Maryland south to the Florida Peninsula; the evidence for this massive destruction lies buried deep below the modern Everglades (discussed later in this book). The catastrophism that resulted from this impact laid the groundwork for the structural features that have shaped the various basins of the Okeechobean Sea over its history and, also in a muted form, the surface of the modern Everglades. Besides encompassing buried oceans and well-preserved marine ecosystems, the Everglades area has now been shown to have a distinct extraterrestrial connection, with its structural geology resulting from the effects of a nearby asteroid impact.

Over its entire history, the Okeechobean Sea basin was flooded extensively ten different times, resulting in the deposition of discrete parcels of marine sediments that were separated from each other by hundreds of thousands of years of terrestrial deposition and erosion. The separate flooding intervals, reflecting patterns of global climate warming, are here referred to as **Subseas** (Petuch, 2004). Each Okeechobean subsea laid down its own characteristic set of marine environments, sediment types, and ecosystems, which allow it to be differentiated from previous or subsequent subseas. After diagenesis (sediment alteration) during sea level lows, these subsea depositional parcels produced the geologic groups, formations, and members that are present within the geographic boundaries of each marine incursion. In this book, we present the description and nomenclatural placement for each of these stratigraphic units, ranging from the formations and members of the early Oligocene to those of the latest Pleistocene. We also use the beautifully-preserved fossils present within these units as chronological markers for determining the ages and structural boundaries of the geologic framework that underlies southern Florida. It is our hope that we will give a new perspective on both the historical geology of southern Florida and the evolution of one of America's most beautiful natural treasures, the Everglades.

Edward J. Petuch, Ph.D. and David P. Berschauer, J.D.

Close-up view of a bed of the large scallop, *Carolinapecten eboreus*, in an exposure of the Pinecrest Member, Tamiami Formation, in Quality Aggregates Pit #6, Sarasota, Sarasota County, Florida. This thick bed of scallops was deposited within the Myakka Lagoon System of late Pliocene (Piacenzian Age) southwestern Florida.

View of Rabbit Key, Ten Thousand Islands, Monroe County, Florida, where the Everglades empties into the Gulf of Mexico. This primordial area resembles the brackish water lagoonal environments of the Pliocene Okeechobean Sea.

CHAPTER 1. Fundamentals of Southern Florida Geology and Paleoceanography

Any in-depth study of the geology of the Everglades area reveals that the local sedimentary framework differs greatly from those seen on the adjacent Floridian Peninsula and the southeastern United States. This unique and distinctive geology is the direct result of almost continuous carbonate sedimentary deposition within a semi-enclosed tropical sea, with each preserved sea floor producing sets of environments and lithofacies that ultimately consolidated into the formations that underlie the Everglades Basin. Over the past hundred years, scientists have become aware of the difficulty in interpreting the geology of such a homogeneous carbonate area. This, coupled with the paucity of exposures within the harsh Everglades swampland, has made southern Florida the last heavily-populated area of the U.S. to be fully explored geologically. Our present understanding of the underpinnings of the Everglades results from the labors, sacrifices, and insights of some of the most dedicated scientists in the history of American geology. These rugged explorers, particularly the early workers, braved the heat, disease-carrying mosquitoes, and dangerous conditions to gather, incrementally, the information that we have assembled here.

Geological and Paleontological Exploration in the Everglades Region.

Considering that southern Florida now houses one of the largest population centers in the United States, it seems amazing that, just a little over a century ago, the area was essentially unexplored. In early 1886, Angelo Heilprin of the Philadelphia Academy of Natural Sciences undertook the first geologic and geographic exploration of the areas around Lake Okeechobee and the Caloosahatchee River. Accompanied by the sponsors of the expedition (Joseph Willcox and Charles H. Brock of Philadelphia), Captain Frank Strobhar, and the cook Moses Natteal, Heilprin investigated the "… mysterious body of water — Okeechobee …" on the schooner "Rambler." This first geologic research was published later in 1886 (first draft; second draft published in 1887) by the Wagner Free Institute of Science of Philadelphia, and Heilprin expounded at length about his adventures in the *Okeechobee Wilderness*. Besides giving the first physiographic description of the northern Everglades region, Heilprin also described new species of living snakes and fishes and some of the more spectacular Everglades fossil shells. This pioneer work excited the scientific community and set the stage for a flurry of geologic and paleontologic research.

For more than a decade, all publications on the geology and paleontology of the Everglades stemmed from a single individual, William Healey Dall of the U.S. Geologic Survey and the Smithsonian Institution (honorary curator). After a preliminary report on the geology of Florida (1887), Dall initiated the *Contributions to the Tertiary Fauna of Florida*, a series of large monographic works that was published between 1890 and 1903 (also by the Wagner Free Institute of Science). Within this same series (1892), Dall also published the first geologic map of Florida, with the Everglades area being left conspicuously blank. Nothing of this scope has ever been attempted since, and the series stands out as one of the great masterpieces of American geology and paleontology. Dall continued publishing works on Floridian geology (a large monograph on the Tampa Oligocene and early Miocene agatized shells) until 1915. Without doubt, William Healey Dall can be considered the "Father of Everglades Geology and Paleontology."

Following Heilprin and Dall's pioneer works, the newly established Florida Geological Survey (formed 1907) immediately began publishing a series of annual reports on exploration of the regional geology. One of the most important of these was G.C. Matson and F.G. Clapp's 1909 preliminary stratigraphic scheme for the Floridian Peninsula (Annual Report 2). Taking into account the lack of natural exposures in the Everglades area, Matson and Clapp's work concentrated on the northern half of the state. In the same year, Samuel Sanford (1909) described and named the surficial limestones of coastal southeastern Florida and the Florida Keys. A decade later, E.H. Sellards (1919) published the first geologic section across the Everglades in the Florida Geological Survey Annual Report 12. As new drainage canals were being dug across the Everglades at that time, Sellards and his coworkers were able to retrieve the first data on what geologic formations existed below the surface peat and muck. All the new information accumulated by the Florida Geological Survey was ultimately synthesized by C.W. Cooke and S. Mossom (1929), who published the first comprehensive work on the geology of the entire state (Annual Report 20). Cooke and Mossom were also the first geologists to traverse the Tamiami Trail. In less than 30 years, the Everglades region had gone from *terra incognita* to being part of the known geologic continuum of the Floridian Peninsula.

As Dall had dominated the 1890s with his prolific works, Wendell C. Mansfield dominated the 1930s with a similar plethora of research and publications. Although his earlier works concentrated on the geology and paleontology of northern Florida, Mansfield began to investigate (in the early 1930s) fossils from newly exposed areas within the Everglades Basin. With his publications in 1931 and 1932 on new fossil mollusks and echinoderms, Mansfield was the first person to discover and name new paleospecies from the southern part of the Everglades. His last work (published posthumously in 1939), which dealt with the stratigraphy of the Everglades Basin, set the stage for all modern studies of the regional geologic framework. Mansfield and Dall, together, named hundreds of important new molluscan stratigraphic index fossils, and their names as the authors of species of fossil shells can be seen throughout this book. In addition to Mansfield, several other important workers described new Everglades molluscan index fossils in the 1930s, including Helen I. Tucker (later Tucker–Rowland), Druid Wilson (1932, 1933) and Maxwell Smith (1936). Most of the discoveries of Tucker and Wilson came from two sources: the construction of the Hoover Dike around Lake Okeechobee following the devastating 1928 hurricane and some of the first commercial quarries in southern Florida near Acline, Lee County.

During the span of World War II, there was a marked drop in the number of publications on Everglades geology. Besides a few scattered papers on new fossil mollusk species by Maxwell Smith and Thomas McGinty (in 1940, before the United States' entry into the war), only a single large work was produced that dealt with the regional geology. This impressive monograph, *The Geology of Florida* by C.W. Cooke (1945), was the culmination and synthesis of all previous research on the geology of the entire State. Although now outdated in content, the book is still a valuable resource for the geology of southern Florida, giving detailed descriptions and illustrations of stratigraphic sections now long lost. Cooke arranged his book geochronologically and illustrated many key index fossils, and we have chosen to use his monograph as a model for this book. Along with Gerald Parker, C.W. Cooke also published the first major study of the groundwater and geologic framework of southern Florida (1944).

With the end of World War II, geologic research in the Everglades region resumed with an intensity not seen since the 1930s. Within a 5-year span, three classic works on the regional geology and paleontology were published: Axel Olsson and Anne Harbison's large monograph on the molluscan paleofauna of St. Petersburg (1953, accompanied by smaller works by Fargo and Pilsbry), Gerald G. Parker's tome on the hydrogeology of the Everglades area (1955; with detailed core logs and deep subsurface data), and Jules DuBar's detailed study on the stratigraphy of the Caloosahatchee River area (1958). These three works represented a quantum leap in our knowledge of the Everglades. The 1960s saw a flurry of commercial development and road building within the Everglades area. Probably the most conspicuous project was the construction of the highway named "Alligator Alley" in Broward and Collier Counties in the early 1960s (now part of Interstate Highway 75, running from Fort Lauderdale to Naples). The preliminary canal dredgings and road-fill excavations produced the first deep transect across the central Everglades. This was a windfall for the regional geologists and paleontologists, allowing them to bulk sample stratigraphic units that had previously been inaccessible or known only from drill cores. New stratigraphic index fossils from "Alligator Alley" and around Lake Okeechobee were reported in a series of publications by Axel Olsson (1967) and Richard E. Petit (1964, 1968) and by Emily Vokes (in the *Tulane Studies in Geology* series; i.e., 1963, 1965, 1966). With the new data from canal excavations around southwestern Florida, Muriel Hunter (1968) was able to offer the first accurate interpretation of the late Miocene and Pliocene stratigraphy of the Everglades region. The same year (1968) also saw the first in-depth study of the geology and origin of the Florida Keys by J.E. Hoffmeister and H.G. Multer. This classic work complemented Hoffmeister and Multer's earlier study (with K.W. Stockman, 1967) on the oölitic limestones of Dade County and the Florida Keys (Miami Formation). Throughout the 1960s, the deepening and channelizing of the Kissimmee River was undertaken by the Army Corps of Engineers. The exposure of sediments in the spoil banks along the Kissimmee River allowed researchers to study, for the first time, the unusual and highly endemic fossil mollusks of that area (the Kissimmee Embayment).

The 1970s marked the beginning of major housing development and road building around the Everglades area, allowing accessibility to many new research sites, particularly along the Kissimmee River in the northern Everglades. At that same time, the demand for fill dirt (primarily for housing construction) produced the economic incentives for large quarries to be excavated all along the southwestern and southern Everglades area. Some of the most important of these included the Ashland Petroleum and Asphalt Corporation (APAC) quarry at Sarasota (originally the Warren Brothers and Richardson pit), the Mule Pen quarry at East Naples (originally a mule breeding site and later the Florida Rock Industries Naples Quarry), and the quarries at Sunniland. Many important new index fossils were described from these sites, including new mollusks (S.C. Hollister, 1971; Emily Vokes, 1975, 1976) and corals (N.E. Weisbord, 1974). All the new data accumulated during the 1950s and 1960s allowed Jules DuBar to build upon his early work on the Caloosahatchee area and create the first workable stratigraphic scheme for the entire Neogene of the Everglades (1974). This work was particularly important because of its descriptions and clarifications of the then poorly known Pleistocene formations. Further research on the Pleistocene and Holocene was undertaken by Enos and Perkins (1977), who published the first detailed study of Quaternary depositional environments in the Everglades and Florida Bay.

The explosion of housing development that started in the 1970s, with increased numbers of landfill quarries and canal digs, escalated even more in southern Florida during the 1980s. Although a disaster for the ecosystems within the Everglades and the peripheral areas, the profusion of new excavations offered geologists many opportunities to study subsurface areas that were previously inaccessible. One of the more important discoveries that resulted from this land development was the uncovering of large, complex Pliocene coral reef systems along southwestern Florida. From data collected in canal digs and quarries in that area, J.F. Meeder was the first to describe the Pliocene reef limestones and associated carbonate facies found along the Collier County and Lee County coasts (1980, based upon his previous work published in a geology field trip guide in 1979). Subsequently, the senior author discovered a "mirror-image" system of Pliocene coral reefs under Dade County (the "Bird Road Dig" of Miami, 1986). When the data on the two contemporaneous reef tracts were later combined, it was shown that the east and west coast systems constituted a single, large contiguous coralline feature, the Everglades Pseudoatoll (Petuch, 1987; for details see Chapter 6 in this book).

Resulting from research in the APAC quarry at Sarasota, the senior author published the first reference stratigraphic column for the Pliocene of southern Florida earlier in the decade (Petuch, 1982; reproduced here in Chapter 5). The first attempt at bringing together geological, environmental, and anthropological studies into a single volume was undertaken by the Miami Geological Society in 1984, with the publication of *Environments of South Florida, Past and Present* (Patrick J. Gleason, editor). The hydrogeology of the east coast Pliocene reef tracts was studied extensively from 1985 to 1986, and the data were subsequently published by L.J. Swayze and W.L. Miller in 1987 (U.S. Geological Survey, Water Resources Division). Thomas M. Scott (Florida Geological Survey) published his large monographic work on the Hawthorn Group of Florida one year later. This pioneer work elucidated the underpinnings of the Everglades basin and set the stage for all subsequent research on the Miocene and early Pliocene of the Everglades area. Paleontological discoveries within the Everglades area continued to be reported throughout the late 1980s, with works by Juan Parodiz (1988, on new mollusks) and the senior author (Petuch, 1987, 1988, 1989; new mollusks).

The 1990s saw a continuation of the intense paleontological research that started in the previous decade. Working at several new fossil sites, the senior author was able to describe over 400 new gastropods (Miocene to Pleistocene; Petuch, 1990, 1991, 1994, 1998; Oligocene, 1997). Similarly, several other workers added important new molluscan index fossils to the Everglades literature, including William Lyons (1991), Juan Parodiz and Jay Tripp (1997), and Geerat J. Vermeij and Emily Vokes (1997). The first integrated geochronological study of a southern Floridian Plio–Pleistocene section (APAC quarry, Sarasota) was published by Douglas Jones *et al.* in 1991, which complimented and supported William Lyons' reanalysis of the regional Plio–Pleistocene stratigraphy (1991). A collection of review papers, giving a broad overview of Floridian geology, was published as *The Geology of Florida* in 1997 (edited by Anthony Randazzo and Douglas Jones of the University of Florida). The volume contains contributions from several geologists and paleontologists who have made important advances in our knowledge of the Everglades region, including Walter Schmidt (Florida State Geologist), Thomas Scott (Assistant State Geologist), and Bruce J. MacFadden. All of the stratigraphic studies published over the past 100 years were summarized and discussed, in a single volume, by Petuch and Roberts in 2007, in a comprehensive text entitled *Geology of the Everglades and Adjacent Areas*. This formed the foundation for the overview of Floridian stratigraphy and paleoceanography that we present here.

The Eocene Chesapeake Bay Asteroid Impact and Its Effects on Florida.

The end of the Eocene Epoch has long been known to have been a time of catastrophic environmental conditions and an accompanying mass extinction, now known to have been caused by an asteroid impact (Ganapathy, 1982; Maurasse and Glass, 1976; Sanfilipo *et al.*, 1985). Based on the chemical composition and distribution of Eocene microtektites in the Caribbean Sea and Gulf of Mexico (Glass and Zwart, 1979), it had been determined that the asteroid impact occurred somewhere along the eastern U.S. Finally, in the late 1990s, the impact site was discovered at the mouth of Chesapeake Bay (Poag, 1999; Powars and Bruce, 1999; Powars, 2000; Poag *et al.*, 2004). When the force of the impact and the size of the Chesapeake Bay Crater were finally reconstructed, it was found that the resultant series of catastrophic events affected the entire Eastern Seaboard, including the Ocala Bank of central Florida and the Florida Platform.

The asteroid that produced the Chesapeake Bay Crater, as impressive as it was judging from the size of its astrobleme, was actually part of an even larger object. Entering the Earth's atmosphere over the North Pole, this asteroid broke into two large fragments and innumerable smaller pieces. The largest fragment (13 kms/8 miles in diameter) hit the continental shelf off Virginia (the present-day mouth of Chesapeake Bay), whereas the next largest fragment hit Popigai, Siberia (Bottomley *et al.*, 1997). The debris field from the American fragment, alone, extended from Chesapeake Bay to off southern Argentina, with a width of over 300 km. The entire eastern U.S. was peppered with millions of fragments, both from the shattered asteroid and from the impact ejecta debris cloud. The main crater, itself, was over 80 km in diameter at impact. The disintegrating asteroid fragment produced not only an initial blast-splash of over 40 km in height, but also tremendous steam explosions and a wide zone of concentric fractures. Today, the resultant impact structure is over 128 km in diameter (Poag, 1999) (Figure 1.1). Numerous suspected astroblemes in the Carolinas, such as Lake Waccamaw, may prove to have been formed by smaller fragments from the disintegrating larger body.

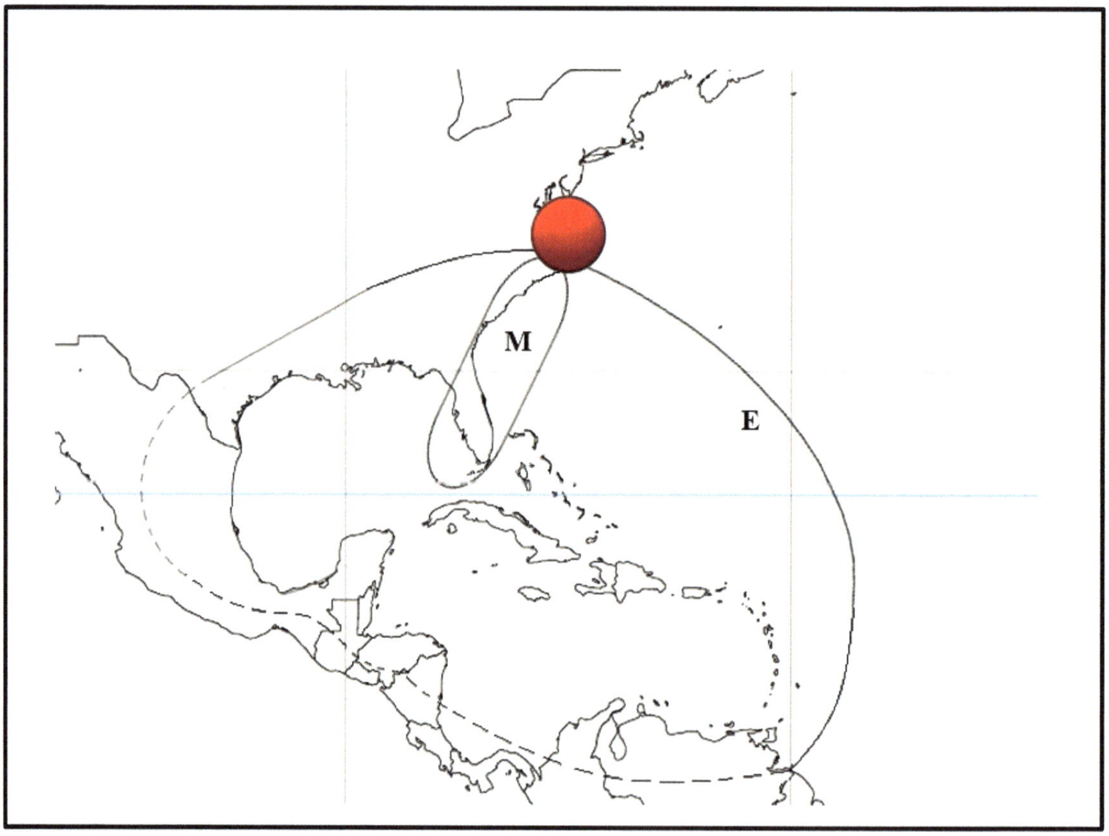

Figure 1.1 Map of the southeastern United States and Central America, showing the approximate areal extent of the debris field produced by the **Chesapeake Bay Asteroid Impact**. The **red circle** represents the location of the impact site at the mouth of Chesapeake Bay; **E**= extent of the ejecta debris blanket and tektite fields; **M**= Zone of Mega-Tsunamis produced by the asteroid impact. The Floridian Peninsula can be seen to have been in the direct line of the largest and most powerful mega-tsunamis, which removed the sea floor of the Florida Platform and produced the **Everglades Unconformity**. Taken from Poag, 1999.

Although the actual impact event was catastrophic in the extreme, what followed had an even greater effect on southern Florida. Within minutes of striking the shallow sea floor off Virginia, the disintegrating asteroid would have generated a series of mega-tsunamis, with some reaching over 500 m in height (Poag, 1999). These would have traveled down the entire eastern coast of the U.S. and, hours later, would have passed over the Ocala Bank and into the Gulf of Mexico. The force of these giant waves would have scoured the entire Florida Platform, removing most of the unconsolidated seafloor sediments. The shallow slope areas east and southeast of the Ocala Bank would have been the first places to be exposed to the full impact of the mega-tsunamis and would have been particularly vulnerable to scouring. This catastrophic event is preserved under the Everglades in the form of a large unconformity (Figure 1.2). Although known to geologists for over two decades (Merritt *et al.*, 1983: 29, Figure 5; Missimer, 1984: 388), these missing sediments, here referred to as the **Everglades Unconformity**, were originally thought to have resulted from some form of postdepositional erosion. In the area of the unconformity, the entire Ocala Group is missing, with the continuously distributed younger (Oligocene) Suwannee Limestone lying directly on top of the Claiborne-aged Avon Park Formation. Considering that the asteroid impact took place at the end of Ocala deposition, the Everglades Unconformity is chronologically perfectly situated and may be the first record of impact-related catastrophism south of the Chesapeake Bay area (for details see Petuch and Roberts, 2007).

Geophysical studies of the Everglades Unconformity area have recently uncovered two interesting anomalies. These are shown in Figure 1.3, superimposed upon the surface contours of the Eocene formations. The first, and most intriguing, is a small magnetic anomaly located within a valley-like depressional feature under the southern Everglades. This strong magnetic anomaly, surrounded by a steep magnetic gradient, was first shown on a magnetic survey map of the Floridian Peninsula produced by Klitgord *et al.* (1984) in their study of Jurassic transform faults. This isolated, geographically small anomaly, with a field of 100 to 250 nT, stands out strikingly from the surrounding low-magnetic area of only 0 to 10 nT. As can be expected on a thick carbonate platform devoid of igneous rocks, the low-magnetic area extends for over 300 km to the south and west of the magnetic high and over 200 km to the north and east. The presence of a well-documented, lone, high-magnetic anomaly, coincident with a valley-like depression in the Eocene rocks, is indeed a most interesting feature.

The second anomalous geophysical feature associated with the Everglades Unconformity is a pair of large gravity lows, which were also illustrated by Klitgord *et al.* (1984) on a gravity survey map from their Jurassic study (shown here on Figure 1.3). These two lows, with fields of less than –5 mG, appear to be related to "high transmissivity zones" or "cavity zones" that are found in the early and middle Eocene, Paleocene, and Cretaceous rocks of the southern tip of the state. These labyrinthine vug structures (referred to as "boulder zones" by drillers) and interconnected cavities may have formed by subsequent dissolution enlargement of brecciated limestone. With the exception of the Everglades area, the older Eocene formations (Oldsmar and

Avon Park), across the State, characteristically contain gypsiferous limestones and anhydrite beds (diagenetically produced). These anhydrites are especially well developed under the central and northern parts of the peninsula. Under the Everglades, there is a distinct shift in lithofacies, with the anhydrite beds being replaced by fractured and vuggy dolostone–limestone facies (Merritt *et al.*, 1983; Puri and Winston, 1974). These "boulder zone" fracture systems follow curving, almost concentric, patterns and house migratory petroleum reserves. The main oil-producing fields on peninsular Florida, the Corkscrew and Sunniland fields, correspond to these curving deep fractures and fault systems along the western and southwestern sides of the Everglades (Applegate, 1986; McCaslin, 1986). Recent investigations on deep-injection well possibilities by private engineering firms (unpublished; personal communication) have shown that these curving fracture systems may also extend around the eastern side of the Everglades (corresponding to the gravity low shown on Figure 1.3). The coincidence of the Everglades Unconformity and the underlying geophysical and geologic anomalies is compelling. These data lead to several possible scenarios, all of which, unfortunately, must remain in the realm of theory for the present time. Considering the new discovery of the Chesapeake Bay asteroid impact, it seems that the two most likely scenarios involve either the effects of the mega-tsunamis themselves or the mega-tsunamis coupled with an impact by a small debris fragment. The first scenario explains the missing seafloor of the Everglades Unconformity and also the gravity low–brecciated limestone patterns (as shattered debris reworked in the tsunami surges). The second scenario, first postulated by the senior author (Petuch, 1987), has the extreme southeastern edge of the Florida Platform being struck by a small impactor, producing a crater- like feature on the seafloor (in 300 m water depth), and shattering the underlying limestones to a depth of 8 km. When the newly discovered mega-tsunamis are also factored into this model, the formation of all the anomalous Everglades features can be combined into a single catastrophic event. Future research within the Everglades may reveal the true nature of the magnetic high, the adjacent gravity lows, and the valley-like structure. Considering the size and density of the Chesapeake impact debris field, geologists should be on the lookout for other small impact structures in the Eocene rocks on the Floridian continental shelf in the Gulf of Mexico and in Georgia and northern Florida.

Stratigraphic Framework and Depositional History of the Everglades Region.

The upper 330 m of the Everglades region, the area covered in this book, encompasses 4 geologic groups, 18 formations, and 23 described members. Geochronologically, these units range from the Upper Eocene (Priabonian Age; Jackson Stage) to the Holocene. Over this 36 million-year time period, there were three post-catatrophic episodes of deposition, with each producing distinct groups of formations. These include:

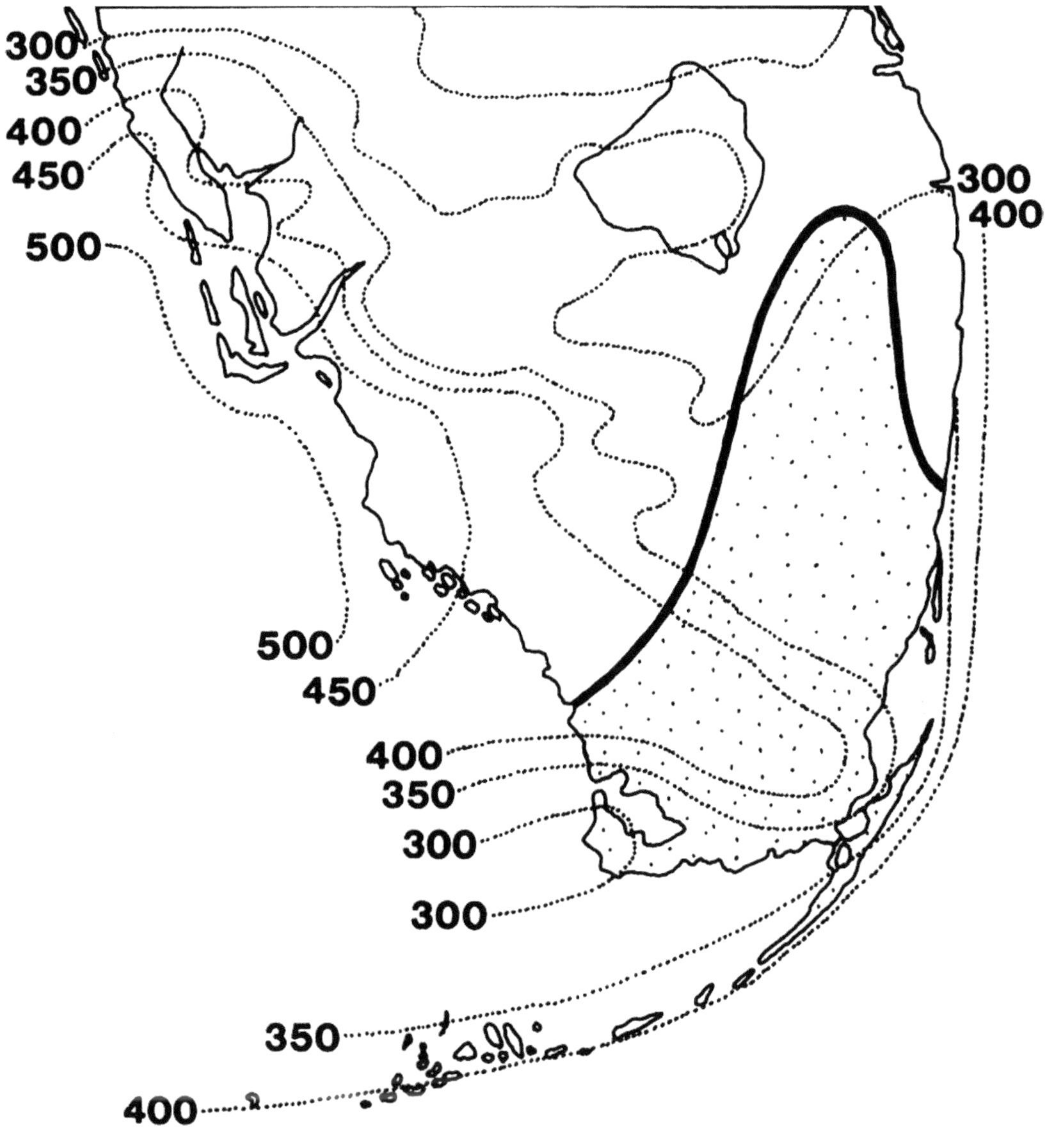

Figure 1.2 Surface contours of the Eocene formations below the Everglades region. The stippled section delineates the area where the late Eocene (Jackson Stage) Ocala Group is missing, and the middle Eocene (Claiborne Stage) Avon Park Formation is exposed. The contours are in meters below mean surface. The area of the missing Ocala Group is referred to here as the **Everglades Unconformity**. (Taken from Merritt, M.I., Myer, F.W., Sonntag, W.H., and Fitzpatrick, D.J., 1983, *Subsurface Storage of Freshwater in South Florida: A Prospectus*, U.S. Geological Survey Water Resources Investigations Report 83–4214, pp. 29–40.)

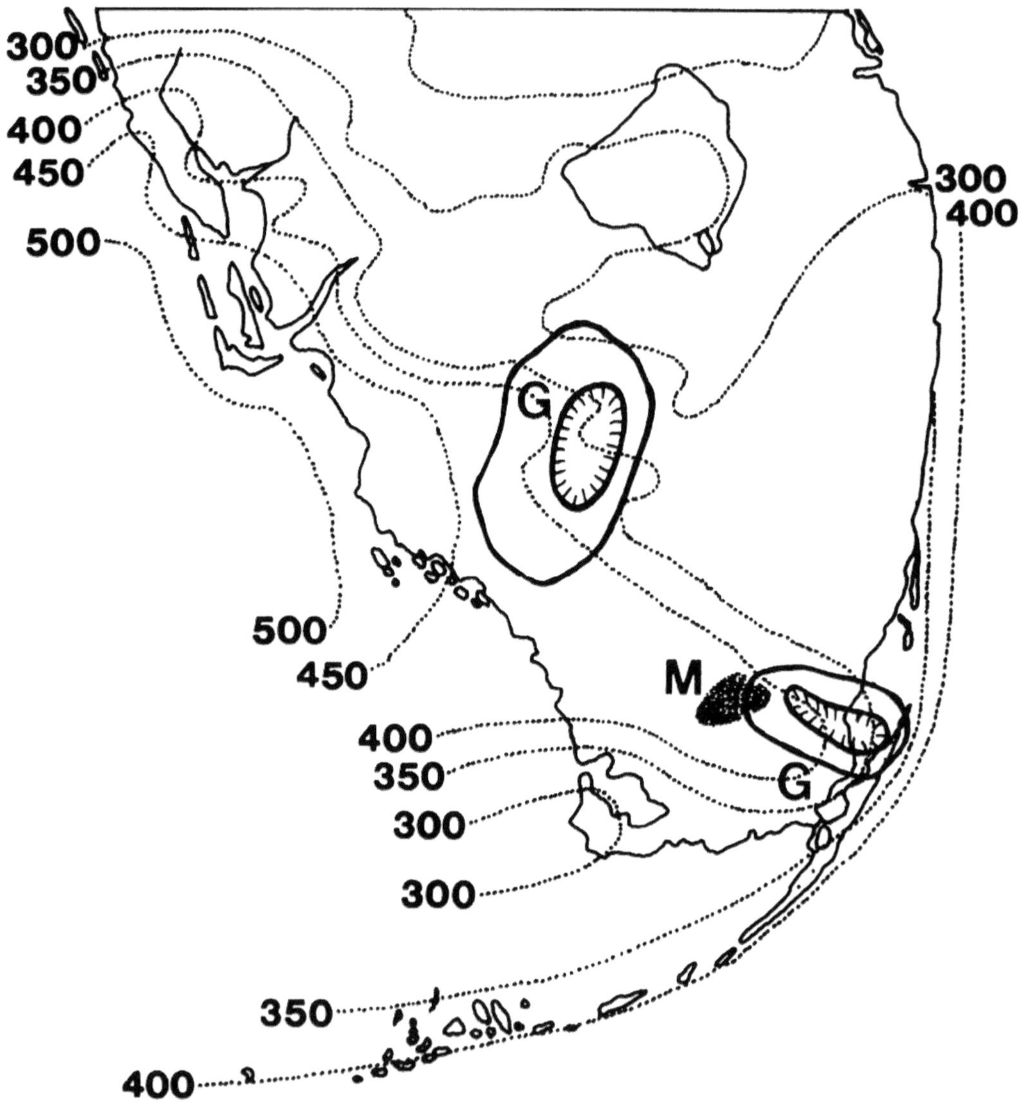

Figure 1.3 Geophysical anomalies below the Everglades area, superimposed upon the surface contours of the Eocene formations. Two gravity lows (G) of –5 mG (milliGals) can be seen to be associated with a magnetic high (M) of 100–250 nT (nanoTeslas). The magnetic anomaly lies in a valley-like feature, and the gravity lows are associated with fracture systems and areas of cavity zones. (Taken from Klitgord, K.D., Popenoe, P., and Schouten, H., 1984, Florida: A Jurassic transform plate boundary, *Journal of Geophysical Research,* 89: 7753–7772.)

1. Carbonate Platform Depositional Period (Ocala and Vicksburg Groups). This period, at roughly 36 million years B.P., represents the incipiency of the deep structural features found below the Everglades area. At that time, Florida consisted of a shallow carbonate bank (the **Ocala Bank**) off the coast of southern Georgia and resembled a smaller version of the Recent Bahamas Banks. The southern edge of the Ocala Bank was in the vicinity of modern Orlando, Orange County, and the northern edge was near Gainesville, Alachua County. To the south, the bank sloped gently to the Straits of Florida, with the mean water depth in the Everglades region being around 200–300 m. Because the Ocala Bank was so far off the coast of Georgia, and separated by the wide **Suwannee Strait**, very little, if any, continental siliciclastic material extended to the southern slope. Here, instead, a constant rain of carbonate fines produced the **Ocala Group** of formations, including the basal **Inglis Formation**, the middle **Williston Formation**, and the upper **Crystal River Formation**. These were deposited on top of the older Eocene (Claiborne Stage) **Avon Park Formation**.The mid-Oligocene (late Rupelian Age) saw a return to high tropical conditions on the Floridian Peninsula, and rising sea levels flooded much of the old Eocene seafloor areas. This tropical inundation led to renewed carbonate deposition, producing the **Suwannee Formation**. This last carbonate platform is here placed within the **Vicksburg Group** of the Mississippi Embayment, primarily because of its lithologic resemblance to the Flint River Formation of Georgia and the Salt Mountain Formation of Alabama. All three of these formations share many index fossil species with the Byram and Mint Springs Formations of Mississippi (classic Vicksburg Group units). As seen in Chapter 2, the Suwannee Limestone is the first continuous, uninterrupted carbonate surface within the Everglades area since the mid-Eocene Avon Park Formation. By the late Oligocene (early Chattian Age), sea levels again regressed, and high tropical carbonate deposition ceased in southern Florida. This marks the end of the Carbonate Platform Episode.

2. Upwelling-Deltaic Depositional Period (Hawthorn Group). From the latest Oligocene (late Chattian Age) to the early Pliocene (Zanclean Age), deposition within the Everglades area changed dramatically. This resulted from two main factors: the establishment of permanent large-scale upwelling systems off western Florida and the accompanying high-productivity marine conditions and plankton blooms (Petuch, 2004: 139), and the southward progradation of giant deltas off the DeSoto Plain highlands region (see Chapter 3). Instead of the Bahamas-like carbonate environments of the Eocene to late Oligocene, the Okeechobean Sea now supported continental terriginous-type deposition, producing thick beds of clay minerals, quartz sand, and phosphorites. The deltas contributed the clay and sand component, whereas the offshore upwelling systems contributed cooler water conditions and abundant phosphate-rich plankton resources. The result was the deposition of the **Hawthorn Group** (Scott, 1988). As presently understood, the Hawthorn Group of the Everglades area comprises four distinct formations: the **Arcadia Formation** (and its **Nocatee** and **Tampa Members**), which extends from the latest Oligocene (late Chattian Age) to the early Serravallian Age of the Miocene; the **Peace River Formation** (and its **Bone Valley Member**), which encompasses the late Serravallian Age and early Tortonian Age of the Miocene; the **Bayshore Formation (**and its **Cocoplum** and **Port Charlotte Members**), which encompasses the late Tortonian and Messinian ages of the Miocene; and the **Murdock Formation** (and its **Jupiter Waterway** and **Sarasota Members**), which was deposited during the early Zanclean Age of the Pliocene. The clay- and phosphorite-rich Hawthorn Group is bounded by transitional facies containing higher percentages of carbonates. This is particularly noticeable in the limestone of the Tampa Member of the Arcadia Formation (basal Hawthorn Group) and the rich carbonate bioclasts of the Sarasota Member of the Murdock Station Formation (uppermost Hawthorn Group). In total, Hawthorn Group deposition lasted for over 20 million years.

down in tropical estuaries at the mouths of rivers along the northern edge of the pseudoatoll lagoon. The facies produced by this diversity of environments are discussed in Chapters 5, 6, and 7 later in this book. The patterns of deposition seen during Tamiami time continued on into the Calabrian Age of the early Pleistocene, resulting in the deposition of the **Caloosahatchee Formation** (with its **Fordville**, **Fort Denaud**, **Bee Branch**, and **Ayers Landing Members**). An extension of the eastern coastal lagoon environments also occurred at this time, resulting in the deposition of the **Nashua Formation** (and its **Fort Drum** and **Rucks Pit Members**). The Caloosahatchee and Nashua depositions are discussed in Chapters 8 and 9. The onset of major and frequent glaciations and sea level drops during the middle and late Pleistocene accelerated the demise of the Everglades Pseudoatoll and ushered in the formation of the modern Everglades. During the sea level drops of glacial stages, the Okeechobean Sea was completely emergent and the Everglades basin contained a series of large freshwater lakes. During warmer interglacial times, the Okeechobean Sea would again reflood, filling in depressional features with biogenic carbonates and erosional fines, and causing the enlargement of older island chains by transported beach sands and mangrove forests. The sequential infilling of the Everglades Pseudoatoll, including emergent times with freshwater lakes and peripheral brackish lagoons, is evident in the depositional patterns of the **Bermont Formation** (with its **Holey Land**, **Okeelanta**, and **Belle Glade Members**) and the **Fort Thompson Formation** (with its **Okaloacoochee** and **Coffee Mill Hammock Members**) (see Chapters 10, 11, and 12). Late Pleistocene beach deposits, coral reefs, and reef-associated oölite banks produced the uppermost formations of the Okeechobee Group: the **Anastasia**, **Key Largo**, and **Miami Formations**, respectively. This last part of the Pseudoatoll Depositional Period is discussed in Chapter 13.

Traditionally, data on southern Floridian stratigraphy had been acquired primarily by using rotary well drilling rig systems. Most recent workers, such as Scott, Missimer, Cunningham, etc., have relied heavily on rock and sediment cores obtained from rotary drill rigs, interpreting the layering and sediment compositions preserved in the cores as the basis for their stratigraphic determinations. The senior author tested the efficacy of core analysis, as opposed to direct physical observation of an exposed geological section, by "ground truthing" an exposed section of the Bermont Formation in a quarry in Palm Beach County (shown here on Figure 1.5). As in most quarries around the Everglades area, this exposed section contained all the fine details of the local stratigraphy, including bedding planes, unconformable surfaces, shifts in lithology, and *in situ* index fossils. At that time, an engineering company was conducting a series of core drilling operations near this exposure, and the senior author requested that one of the cores be drilled in close proximity, so that the contents of the core could be compared directly with the adjacent section. The test core, taken from only 7 m from the edge of this section, failed to reveal the sharply-defined bedding plane surfaces seen in the photograph, instead yielding only a homogenized mass of shell fragments, mud, and sand. The core bore no resemblance at all to the section seen here and demonstrated that rotary drill rigs are essentially

Figure 1.5 Close-up of an exposed section of the Holey Land Member, Bermont Formation, in the Palm Beach Aggregates Pit #6, showing distinct alternating beds composed of unconsolidated mollusk shells and indurated sandy limestones. A test core, taken by a rotary drill only 7 m from the edge of this section, failed to reveal the sharply-defined bedding plane surfaces seen here, instead yielding only a homogenized mass of shell fragments, mud, and sand. The core bore no resemblance at all to the section seen here and showed that rotary drill rigs are essentially useless for studying the fine details of surficial geology. Photo by Anton Oleinik.

useless in studying the fine details of the surficial geology. Only personal observation of exposed sections in de-watered quarries, canals, and excavations can be used to accurately determine the stratigraphy of the Everglades area.

The Okeechobean Sea and Its Subseas.

Throughout this chapter, we have referred to the Okeechobean Sea primarily in the broad sense, as a single body of water covering the Everglades area. Over the past 36 million years, the Okeechobean basin actually has been flooded by 10 separate large-scale marine incursions. Each of these, in turn, had intervals of minor sea level fluctuations superimposed upon the larger transgressive pattern. As will be seen in the following chapters, the larger transgressions resulted in the deposition of the Everglades geologic formations whereas the minor transgressions resulted in the deposition of their members and beds. The intervening regressive intervals varied in time from millions of years (between the large transgressive intervals) to thousands of years (between the smaller transgressive intervals). About one third of the 36 million-year history of the Okeechobean Sea is lost due to multiple extended periods of terrestrial conditions.

As discussed earlier in this chapter, the depositional and chronological subdivisions of the Okeechobean Sea are here referred to as "subseas." Between the time of the Ocala Sea and the mega-tsunamis (latest Eocene) and Recent Florida, we recognize 10 distinct subsea divisions of the Okeechobean Sea. These oceanographic subdivisions, which represent major marine transgressive intervals and extended periods of deposition, were named and described by the senior author (Petuch, 2004) and Petuch and Roberts (2007) and include (from oldest to youngest):

1. **Dade Subsea** (early Oligocene; named for Dade County; see Chapter 2.)

2. **Tampa Subsea** (late Oligocene-earliest Miocene; named for Tampa, Hillsborough County, and the Tampa Member of the Arcadia Formation; see Chapter 3.)

3. **Polk Subsea** (early late Miocene; named for Polk County; see Chapter 4)

4. **Charlotte Subsea** (latest Miocene; named for Charlotte County; see Chapter 4)

5. **Murdock Subsea** (early Pliocene; named for Murdock, Charlotte County, and the Murdock Formation; see Chapter 4)

6. **Tamiami Subsea** (late Pliocene; named for the Tamiami Trail in Dade and Collier Counties, and the Tamiami Formation; see Chapters 5, 6, and 7)

7. **Caloosahatchee Subsea** (early Pleistocene; named for the Caloosahatchee River and the Caloosahatchee Formation; see Chapters 8 and 9)

8. **Loxahatchee Subsea** (early middle Pleistocene; named for Loxahatchee, Palm Beach County; see Chapter 10)

9. **Belle Glade Subsea** (late middle Pleistocene; named for Belle Glade, Palm Beach County; see Chapter 11)

10. **Lake Worth Subsea** (late Pleistocene; named for Lake Worth, Palm Beach County; see Chapter 12)

In each of the following chapters, the prominent geomorphological features of each of these subseas are described, along with their paleoceanography and depositional environments. Iconographies of prominent marine organisms, including over 800 species of corals, mollusks, echinoids, and vertebrates are given at the end of each chapter or section. These selected fossil organisms can be used as stratigraphic and geochronological index fossils or for paleoecological reconstructions of ancient environments. See Petuch and Drolshagen (2011) for a review of the fossil gastropod families **Strombidae**, **Triviidae**, and **Ovulidae**; Petuch, Drolshagen, and Herndl (2015) for a review of the fossil **Conidae**; Petuch, Myers, and Berschauer (2015) for a review of the fossil **Busyconidae**; Petuch, Berschauer, and Myers (2018) for a review of the fossil **Cypraeidae**; and Petuch (1994) for an iconography of the entire marine macrogastropod fauna (Late Miocene to Late Pleistocene, including 1,100 species in 237 genera and 56 families).

CHAPTER 2. The Dade Subsea (Suwannee Formation)

Immediately following the late Eocene asteroid impact, the world was plunged into a severe cold time ("nuclear winter"), with expansive ice buildup in the Northern Hemisphere. This led to a major sea level drop, with eustatic lows reaching 200 to 300 m below present sea level. From the latest Eocene (36 million years) until the mid-Rupelian Oligocene (32 million years), southern Florida remained above sea level, with dry subtropical to warm-temperate conditions prevailing. During this 4 million-year terrestrial episode, the Avon Park Formation (in the Everglades) and the Crystal River Formation (elsewhere in Florida) were exposed as the land surface across the entire Florida Platform. In the area of the Everglades Unconformity, groundwater penetrated the underlying fractures, dissolving the smaller limestone fragments and enlarging the spaces between brecciated blocks. Over time, this dissolution caused the entire southern Everglades area to slowly collapse, forming a shallow depressional feature. This dissolutional collapse feature may also have formed as an enlarged version of the older late Eocene valley structure. Whatever its means of formation, the southern Everglades depression is an obvious feature and is readily demonstrated in the stratigraphic columns of Puri and Winston (1974). In their data, the top of the younger Oligocene Suwannee Formation is seen to conform to a 100 m-deep depression.

By the beginning of Rupelian time, the worldwide climate began to warm and sea levels rose almost to their late Eocene levels. A shallow sea, the **Dade Subsea** (named for Dade County, Florida; Figure 2.1), now flooded the southern Florida Platform and produced depositional environments that resulted in the sediments of the Suwannee Formation (Everglades component). The slightly lower sea levels of the early Oligocene allowed the remnants of the Ocala Bank to remain above water, forming a large, low-relief land mass named **Orange Island** by Vaughan (1919). Because of lower sea levels, the **Suwannee Strait** was now much shallower than during the Eocene and contained extensive coral reef systems. These reefs extended northward, into southern Alabama (the Salt Mountain Formation) and southern Georgia (the Flint River Formation), and southward, forming a large reef tract (the **Pasco Reef System**) along the western side of Orange Island. The Pasco Reefs housed amazingly rich and endemic coral and molluscan faunas, many of which are extinct in the Atlantic Ocean but still living in the South Pacific. The entire area more closely resembled present-day northern Australia than it did the modern Caribbean (Petuch, 1997). The area of southern Alabama, southern Georgia, Suwannee Strait, and northern Orange Island was included within the Bainbridge Subsea of the Choctaw Sea (see Petuch, 2004: 2–5). Only the area south of Orange Island constituted the Dade Subsea of the Okeechobean Sea.

At the extreme southeastern end of the Florida Platform, two arc-shaped carbonate banks had formed around the higher edges of the southern Everglades depression. These two shallowly-submerged structures created the template for all subsequent

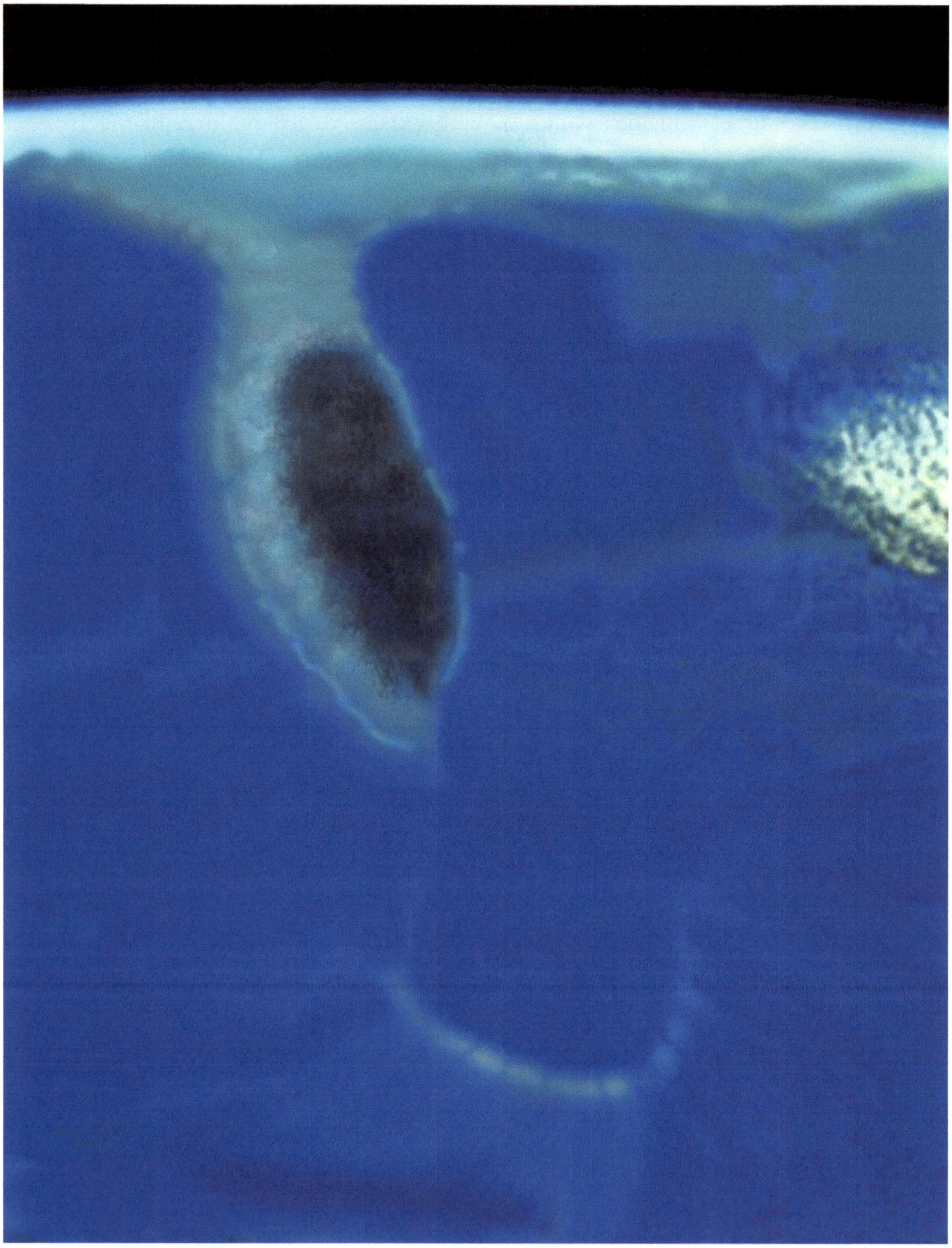

Figure 2.1 Simulated space shuttle image (altitude: 100 mi) of the Floridian Peninsula during the Rupelian Oligocene, showing the probable appearance of the **Dade Subsea** of the Okeechobean Sea. The large, low **Orange Island** had now formed on the surface of the old Ocala Bank and was separated from the North American mainland by the wide **Suwannee Strait** to the north. The extensive coral reef complexes of the **Pasco Reef System** grew along the western side of Orange Island and numerous smaller reef systems were also present within the Suwannee Strait.

reefal deposition in the Everglades area. The Suwannee Formation sediments from these bank areas are composed of unconsolidated calcarenites and thick beds of foraminiferal tests (Missimer, 1984: figure 6). Separating the depressional feature and its arc-shaped banks from Orange Island was a wide, deep channel. By the end of Suwannee time (early Chattian Oligocene), the central depression began to fill with carbonate fines that were eroding off Orange Island and the surrounding reef systems.

The first faint outline of the modern Floridian Peninsula was beginning to emerge. The paleoenvironments contained in the Suwannee Formation were described by the senior author (Petuch, 2004: 57–67). Although known for decades from exposures along the Suwannee River and from quarries in north central Florida, the Suwannee Formation was not formally described until 1936, when Cooke and Mansfield proposed the name "Suwannee Limestone." Because of its commercial importance as a local source of railway ballast and as an aggregate for concrete (Cooke, 1945: 90–91), the formation had been chemically analyzed over a decade before it was formally named (Mossom, 1925). The Suwannee is still quarried as a source of limestone aggregate in Citrus, Hernando, Pasco, and Polk Counties

Lithologic Description of the Suwannee Formation: (as described by Cooke, 1945: 87; condensed here): "Where the Suwannee is unaltered. It consists of a soft granular mass of limy particles, many of them of organic origin. The color is commonly yellow or cream, locally with a pinkish tinge. At many surface exposures all the lime has been leached from it, leaving a porous or massive flint, which is recognizable as Suwannee by the presence of molds of the common Suwannee echinoid, *Cassidulus gouldii* [authors' note: now *Rhyncholampas gouldii*] … . Chemical analyses show that the Suwannee limestone contains about 91–98% calcium carbonate and that the chief impurity is silica." The lithology of the Suwannee Formation in the Everglades was best summarized by Missimer (1984: 388–389) (condensed here): "Lithologically, the Suwannee is predominantly a light-colored limestone, which commonly contains soft calcarenite consisting of foraminifera tests. It does contain some unlithified lime muds and well-indurated dolomites. The dolomites are commonly hard, thin, and interbedded with limestones. Some unlithified dolosilts occur in the middle part of the stratigraphic section …" The depositional area of the Suwannee conforms to the paleogeography of the Dade Subsea. In northern and western peninsular Florida, the formation follows the contours of the coral reef systems and carbonate lagoons of the Suwannee Strait and its reef systems. These limestones often contain beautifully preserved siliceous pseudomorphs (agatized specimens) of an especially rich mollusk and coral fauna. Specimens of these Suwannee agatized fossils have been found as far south as DeSoto County. South of there, these macrofossils are replaced by microfossils, mostly foraminifera, reflecting the deep water conditions of the Oligocene Everglades. Stratigraphically, the Suwannee appears to be more complex than originally thought, containing at least three distinct units. Future research may show that these constitute members yet to be described.

The Suwannee Formation underlies the entire area that is covered by this book. As mentioned earlier, the formation has a conformable contact with the underlying late Eocene Ocala Group throughout the region, with the exception of the Everglades area. There, the Suwannee lies unconformably on the middle Eocene Avon Park Formation. The thickness of the Suwannee in southern Florida is variable, with a maximum of 180 m under Lee County and a minimum of 18 m under Dade County. The general surface contours of the Suwannee show a gradual dip to the south, being approximately 100 m below the surface in DeSoto County, 250 m in Palm Beach County, 210 m in Lee County, 300 m in southern Palm Beach County and Broward County, and over 400 m in the depressional feature in western Dade and northernmost Monroe Counties. South of this feature, in Monroe County (including Florida Bay and the Florida Keys), the Suwannee Formation comes closer to the surface, at around 200 m below the northern Keys (Puri and Winston, 1974).

Stratotype: (from Cooke, 1945: 86): " … The name "Suwannee Limestone" was proposed by Cooke and Mansfield (1936) for a yellowish limestone typically exposed along the Suwannee River, from Ellaville almost to White Springs." A typical section was illustrated by Cooke (1945: 88, Figure 12; a photograph by Herman Gunter) from the left bank of the Suwannee River "… about half a mile above the old bridge on the road from White Springs to Lake City" (now U.S. Highway 41), Hamilton County.

Age and Correlative Units. Based on the presence of many classic Mississippi Vicksburg Group fossil mollusks, the Suwannee Formation is now known to span the period from the early middle Rupelian Oligocene to the earliest Chattian Oligocene, a range of over 5 million years. Some of the more important of these classic Mississippi Oligocene fossils found in the Suwannee include the gastropods *Clavolithes vicksburgensis*, *Conorbis porcellanus*, *Ficus mississippiensis*, *Torcula mississippiensis*, *Talityphis mississippiensis*, and *Terebrellina divisura*. These typical Vicksburg index fossils, and many others, were reported from the Suwannee and illustrated for the first time by Petuch (1997; 2004: 57–67) and Petuch and Roberts (2007). The Suwannee Formation of the Everglades region is exactly correlative with the Byram and Mint Springs Formations of the Vicksburg Group of Mississippi, the Flint River Formation of southern Georgia, and the Salt Mountain Formation of Alabama.

Suwannee Formation Index Fossils. Along with the Mississippi Vicksburg Group species, the Suwannee molluscan fauna contains a large endemic component that can be used as index fossils to demarcate the formational boundaries. The Suwannee fauna also shares many species with the Flint River Formation of Georgia. These all appear to have been associated with the Rupelian Oligocene coral reef systems of the Suwannee Strait area. Most of these shared taxa were previously described and illustrated by Petuch (1997; 2004) and Petuch and Roberts (2007). Of particular importance as an index fossil for the Suwannee Formation, as pointed out by Cooke

(1945), is the echinoid *Rhyncholampas gouldii* (Figure 2.4E) (originally placed in the genus *Cassidulus*). This small echinoid is the most conspicuous and widespread fossil in the Suwannee Formation. In the reefal facies of the Suwannee, a large fauna of corals can also be used to delineate the formational boundaries. Many of these corals are shared with the Flint River Formation of Georgia and the Salt Mountain Formation of Alabama. Others are shared with the Antigua Formation of the Lesser Antilles (Caribbean Oligocene). Many of the principal molluscan and scleractinian index fossil for the Suwannee Formation are shown on the plates in the following iconography.

Figure 2.2 Close-up of an agatized rib fragment from the Oligocene Dugong, *Metaxytherium albifontanum* Velez-Juarbe and Domning, 2014, a rare species known from only a few localities in South Carolina and Florida. This 112 mm specimen, from the Terramar Pit in rural Polk County, Florida represents the southernmost record for this species.

Figure 2.3 Close-up of a slab of silicified limestone from the Suwannee Formation, in the Terramar Pit, rural Polk County, Florida, showing three imbedded specimens of the small cowrie *Cypraeorbis kendrewi* (orange-red smooth spherules on the lower left). All of the fossils shown on Figures 2.4-2.9 were collected at this locality.

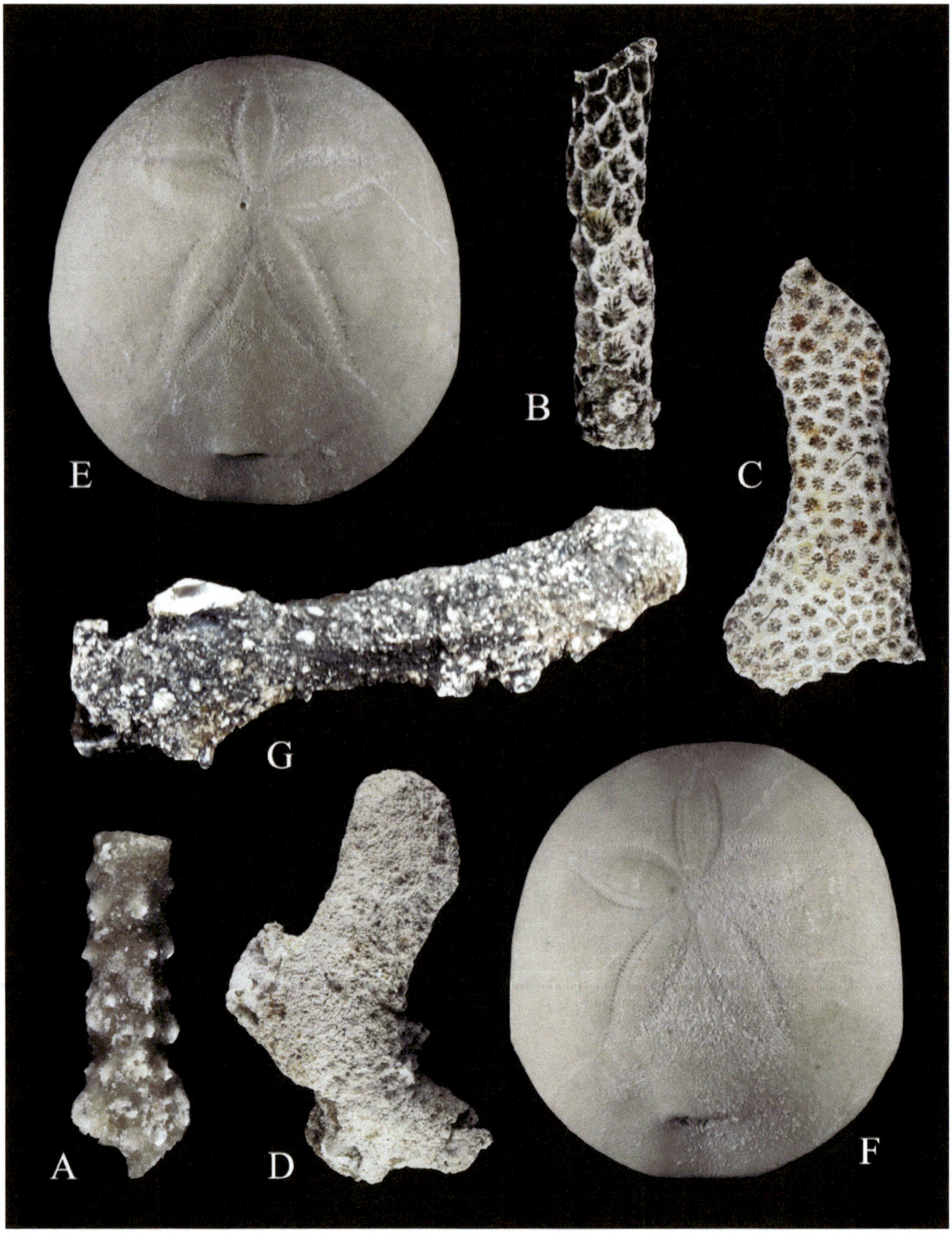

Figure 2.4 Corals and Echinoids of the Suwannee Formation.
A= *Acropora saludensis* Vaughan, 1919, length 22 mm; **B**= *Stylophora minutissima* Vaughan, 1900, length 19 mm; **C**= *Stylophora ponderosa* Vaughan, 1900, length 33 mm; **D**= *Goniopora decaturensis* Vaughan, 1919, length 34 mm; **E**= *Rhyncholampas gouldii* (Bouve, 1846), length 42 mm; **F**= *Rhyncholampas ericsonii* (Fischer, 1951), length 43 mm (this species has a higher and more domed test than does the flatter and more compressed *R. gouldii*); **G**= *Actinactis alabamensis* Vaughan, 1900, length 73 mm. All specimens are agatized pseudomorphs.

Figure 2.5 Mollusks of the Suwannee Formation.
A= *Solenosteira suwanneensis* Petuch, 1997, length 27 mm; **B, C**= *Cypraeorbis kendrewi* Petuch, 1997, length 26 mm; **D**= *Atlanticonus kendrewi* (Petuch, 1997), length 33 mm; **E**= *Pterynotus propeposti* (Mansfield, 1937), length 35 mm; **F**= *Falsilyria kendrewi* Petuch, 1997, length 36 mm; **G**= *Vasum suwanneensis* Petuch, 1997, length 45 mm; **H**= *Calliostoma silicatum* Mansfield, 1937, width 14 mm; **I**= *Pachycrommium dalli* Petuch, 1997, length 26 mm; **J**= *Orthaulax hernandoensis* Mansfield, 1937, length 70 mm; **K**= *Campanile blackwaterense* (Mansfield, 1937), length 43 mm. All specimens are agatized pseudomorphs.

Figure 2.6 Mollusks of the Suwannee Formation.
A= *Cestumcerithium pascoensis* (Mansfield, 1937), length 37 mm; **B**= *Campanile hernandoense* (Mansfield, 1937), length 49 mm; **C**= *Suwannescapha lindae* Petuch, 1997, length 18 mm; **D**= *Prunum sandrae* Petuch, 1997, length 18 mm; **E**= *Amauropsis mansfieldi* Petuch, 1997, length 28 mm; **F**= *Pleurofusia dowlingi* Petuch, 1997, length 21 mm; **G**= *Omogymna brooksvillensis* (Mansfield, 1937), 20 mm; **H**= *Turbinella suwanneensis* (Mansfield, 1937), length 107 mm; **I**= *Cestumcerithium brooksvillensis* (Mansfield, 1937), length 24 mm; **J**= *Pyrazisinus kendrewi* Petuch, 1997, length 30 mm. All specimens are agatized pseudomorphs.

Figure 2.7 Suwannee Formation Mollusks Shared with the Flint River Formation, Vicksburg Group.
A= *Stephanoconus vaughani* (Dall, 1916), length 20 mm; **B=** *Cerithioclava eutextile* (Dall, 1916), length 57 mm; **C=** *Conasprelloides cookei* (Dall, 1916), length 38 mm; **D=** *Falsilyria mansfieldi* (Dall, 1916), 32 mm; **E=** *Apicula bowenae* (Mansfield, 1937), length 39 mm; **F=** *Ampullinopsis flintensis* (Mansfield, 1937), length 32 mm; **G=** *Cassis flintensis* Mansfield, 1940, length 58 mm; **H=** *Lucina perovata* (Dall, 1916), length 32 mm. All specimens are agatized pseudomorphs.

Figure 2.8 Suwannee Formation Mollusks Shared with the Mint Springs Formation, Vicksburg Group.
A= *Ficus mississippiensis* Conrad, 1848, length 23 mm; **B**= *Semicassis caelatura* (Conrad, 1848), length 29 mm; **C**= *Cypraeorbis sphaeroides* (Conrad, 1848), length 27 mm; **D**= *Ostrea vicksburgensis* Conrad, 1848, length 73 mm; **E**= *Laevityphis mississippiensis* (Conrad, 1848), length 24 mm; **F**= *Myristica crassicornuta* (Conrad, 1848), length 49 mm; **G**= *Gradiconus alveatus* (Conrad, 1848), length 23 mm; **H**= *Takia cookei* (E. Vokes, 1975), length 27 mm. All specimens are agatized pseudomorphs.

Figure 2.9 Suwannee Formation Mollusks Shared with the Mint Springs Formation, Vicksburg Group.

A= *Conorbis porcellanus* (Conrad, 1848), length 36 mm; **B**= *Torcula mississippiensis* (Conrad, 1848), length 51 mm; **C**= *Terebrellina divisura* (Conrad, 1848), length 54 mm; **D**= *Fusimitra conquisita* (Conrad, 1848), length 35 mm; **E**= *Conomitra vicksburgensis* (Conrad, 1848), length 17 mm; **F**= *Antillophos vicksburgensis* (Aldrich, 1885), length 22 mm; **G**= *Chicoreus stetopus* (de Gregorio, 1890), length 32 mm; **H**= *Clavolithes vicksburgensis* (Conrad, 1848), length, length 89 mm. All specimens are agatized pseudomorphs.

Cross-section of a 135 mm branch of an agatized fossil *Porites* coral from the Aquitanian Miocene Tampa Member of the Arcadia Formation, showing the beautiful red and blue agate within the interior of the hollow fragment. Polished pieces of this semi-precious agate still command high prices and are frequently used in jewelry. The Tampa Bay Agatized Coral is the official State Rock of Florida.

CHAPTER 3. The Tampa Subsea (Arcadia Formation)

During the middle Oligocene (early Chattian Age), the world climate again cooled, and this resulted in a sea level drop that exposed the entire Dade Subsea, Orange Island, and the Suwannee Strait area to dry land conditions. After over 1 million years of terrestrial environments, the Okeechobean Sea again reflooded the Everglades area, this time as the **Tampa Subsea** (named for Tampa, Hillsborough County) (Figure 3.1). During the earliest Aquitanian Miocene, Orange Island had enlarged both northward and westward (to present-day DeSoto County), with a large tropical carbonate lagoon system having formed along the western side. Fringing the western edge of this lagoon system were the principal geomorphologic feature of the Tampa Subsea: the **Tampa Reef Tract**, a zonated coral reef system, and its associated **Tampa Archipelago**, a linear chain of small coral cays resembling the Recent Bahamian Islands. The first extensive permanent upwelling systems began at the end of Aquitanian time, with phosphate-rich plankton blooms being washed into the Okeechobean Sea basin.

To the south of Orange Island and the Tampa Reef Tract, the Everglades Basin still remained a deepwater area, with depths of over 250 m along the eastern side. Offshore oyster banks that had formed Collier and Dade Counties also expanded during the Aquitanian Miocene, becoming wider and extending farther northward (referred to as the **Collier Bank** and **Dade Bank**; Petuch and Roberts, 2007). In the north, the Suwannee Strait and adjacent areas in Georgia had begun to fill in, becoming shallower and narrower. Extensive mangrove forests were established over much of that area during this time, contributing to the increased shoaling. The Tampa Reef Tract and the Everglades Basin served as the centers of deposition for the Nocatee and Tampa Members of the Arcadia Formation (Scott, 1988). These basal members represent the beginning of the Hawthorn Group. First formally proposed by Scott (1988), the Arcadia Formation exhibits complex depositional patterns, both geographically and chronologically, and spans the entire time frame of the Tampa Subsea. The Arcadia Formation contains the basal Nocatee, Tampa, and lower Arcadia Members and an upper Type Arcadia.

Lithologic Description of the Arcadia Formation. As described by Scott (1988: 56–58; condensed here): "... The Arcadia Formation, with the exception of the Nocatee Member, consists predominantly of limestone and dolostone containing varying amounts of quartz sand, clay, and phosphate grains. Thin beds of quartz sand and clay often are present scattered throughout the section. These thin sands and clays are generally very calcareous or dolomitic and phosphatic ... Dolomite is generally the most abundant carbonate component of the Arcadia Formation, except in the Tampa Member. Limestone is common and occasionally is the dominant carbonate type. The dolostones are quartz sandy to very porous (moldic porosity) and micro-to-fine crystalline. The dolostones range in colorfrom yellowish-gray to light olive-gray. The phosphate grain content is highly variable, ranging up to 25% but is more commonly in the 10% range ... the limestones are typically a wackestone to mudstone with few

Figure 3.1 Simulated space shuttle image (altitude: 100 miles) of the Floridian Peninsula during the early Aquitanian Miocene, showing the probable appearance of the **Tampa Subsea** of the Okeechobean Sea. At this time, Orange Island had expanded in size and a large coral reef system and chain of coral islands, the **Tampa Reef Tract** and the **Tampa Archipelago**, had developed along its western side, along with the **Hillsborough Lagoon System**. Far offshore to the south, the **Collier Bank** and **Dade Bank** had developed from near-surface oyster beds.

beds of packstone. They range in color from white to yellowish-gray … fossils are generally present only as molds in the carbonate rocks … Clay beds occur sporadically in the Arcadia Formation … They are thin, generally less than 5-feet thick, and of limited areal extent … color of the clay ranges from yellowish-gray to light olive-gray … smectite, illite, palygorskite, and sepiolite comprise the clay mineral suite … chert is also sporadically present in the Arcadia Formation in the updip areas … in many instances, the chert appears to be silicified clays and dolosilts." The Arcadia Formation underlies the entire area covered by this book and follows the general contours of the underlying Suwannee Formation. As pointed out by Scott (1988: 60), the top of the Arcadia ranges from 146 m below surface in Monroe County to 30 m below surface in Polk County. In the area of the Everglades Unconformity and depression feature (Palm Beach and Martin Counties), the top of the Arcadia is greater than 250 m below surface.

Stratotype. From Scott (1988: 56): "… The Arcadia Formation is named for the town of Arcadia in DeSoto County, Florida. The type section is located in core W-12050, Hogan #1, DeSoto County (SE ¼, NW ¼, Sec. 16, T. 38 S., R. 26 E.), surface elevation 19 m, drilled in 1973 by the Florida Geological Survey. The type Arcadia Formation occurs between –30 m MSL to –159 m MSL."

Age and Correlative Units. Geochronologically, the Arcadia Formation ranges from the earliest Aquitanian Miocene to the Serravallian Miocene, with a 4 million-year hiatus between the upper and lower members. In Florida, the Tampa Member and Nocatee Member, and the lower Arcadia beds, are exact equivalents of the Penney Farms Formation and the Chattahoochee and St. Marks Formations of the Panhandle. The upper Type Arcadia is the exact equivalent of the Marks Head Formation of northern Florida, the Torreya Formation of the eastern Panhandle, and the Chipola and Oak Grove Formations of the Alum Bluff Group.

Nocatee Member, Arcadia Formation. This member from the lowest part of the Arcadia Formation was deposited along the southeastern corner of Orange Island and within the eastern section of the Hillsborough Lagoon System. The high quartz sand and clay content of the Nocatee was probably the result of a series of rivers discharging into the headwaters of the lagoon. These rivers would have transported sand and clay from the higher elevations on Orange Island downslope and into the area of the Tampa Reef System. These river-borne siliciclastic sediments, mixed with carbonate erosional fines that derived from the coral reefs, would have poured into the deep Okeechobean basin and deposited the lower beds of the undifferentiated Arcadia. The sediments of the Nocatee Member demonstrate that large Orange Island river systems were draining into the Tampa Subsea during the Aquitanian and Burdigalian Miocene. These same rivers, later in the Miocene, may have been the source of the great delta systems (see Chapter 4).

Lithologic Description and Areal Distribution. From Scott (1988: 73; condensed here): "… The Nocatee is a predominantly siliciclastic unit in the type core. This is a noticeable change from the remainder of the Arcadia Formation, including the Tampa Member, which are predominantly carbonates with variable percentages of included siliciclastics. The quartz sands in the Nocatee are typically coarse grained … and range in color from white to light olive-gray. Phosphate grain content is quite variable. In the type core, phosphate grain content is generally low (1 to 3%) with scattered beds with greater concentrations (up to 10%) … in the Nocatee Member in other cores, phosphate grains are more common, averaging 7 to 8% … Clay beds are quite common and are variably quartz sandy, silty, phosphatic, and calcareous to dolomitic. The colors characteristically range from yellowish-gray to light olive-gray and olive-gray. Limited x-ray data suggest that the characteristic clay mineral present is smectite, with palygorskite common. Illite and sepiolite are also present … Limestone and dolostones are both present in this member … The limestones are generally fine grained, soft to hard, quartz sandy and phosphatic … Colors of the limestones vary from white to yellowish-gray to olive-gray … The limestones are usually wackestones with varying degrees of recrystallization and cementation … The dolostones are quartz sandy, phosphatic, soft to hard. And micro-to-very finely crystalline … colors range from yellowish-gray to light gray, light olive-gray, and grayish-brown … Fossils are often present in the Nocatee, most often as molds." The Nocatee Member is of limited areal extent (Figure 2.8), having been identified only in Polk, Hardee, DeSoto, Charlotte, Lee, Manatee, Hillsborough, Sarasota, and westernmost Glades and Highlands Counties. The top of the Nocatee ranges in depth from –24.5 m MSL in Polk County to –195 m MSL in Charlotte County."

Stratotype. From Scott (1988: 73): "… The Nocatee Member is named for the town of Nocatee in central DeSoto County, Florida. The type core is W-12050, Hogan #1, located in the SE ¼, NW ¼, Sec. 16, T. 38 S., R. 26 E., DeSoto County, Florida, with a surface elevation of 19 m. The type Nocatee occurs between –89.5 m MSL and –15.8 m MSL. The type core was drilled by the Florida Geologic Survey."

Tampa Member, Arcadia Formation. Because of its lithologic differences with, and chronological isolation from, the main Type Arcadia beds (late Chattian–Aquitanian as opposed to mid-Burdigalian–Serravallian), the senior author (Petuch, 2004: 16) elevated the Tampa to full formational status, retaining the nomenclature of the early workers such as Dall (1890) and Heilprin (1886). In their 2007 book, Petuch and Roberts decided to return to Scott's stratigraphic scheme and considered the Tampa to be only a member of the Arcadia Formation; a subordinate stratigraphic position that we support in this book. Future workers may wish to revisit the lithostratigraphic and nomenclatural status of this distinctive and historically important unit.

Surface outcrops of the Tampa Member at Tampa Bay have been known for over 100 years and were first studied by Angelo Heilprin during his investigations of the Okeechobee Wilderness (1886). Based on Heilprin's research on the "Silex Beds," L.C. Johnson (1883) proposed the name Tampa Formation for these fossiliferous outcrops. Dall (1892) later elevated the Tampa to group status, bringing together the

Chipola Formation and "Alum Bluff Beds" (now known to be Chipola outcrops on the Apalachicola River) of the Florida Panhandle and the Tampa "Silex Beds" at Tampa Bay. Finally, Matson and Clapp (1909) formally described the "Tampa Formation" and restricted the use of the name to only the beds exposed in the vicinity of Tampa. More recently, King and Wright (1979) redescribed the Tampa Formation in detail and designated a stratotype (at Ballast Point, Hillsborough County). Later, Scott (1988: 73) reduced the status from formation to member, stating that " ... the reduction is necessary due to the limited areal extent of the Tampa and its interfingering, gradational nature with part of the Arcadia Formation."

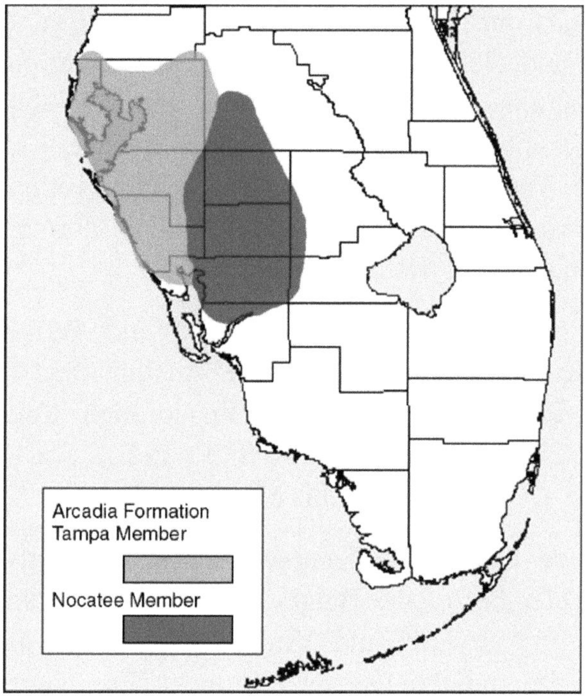

Figure 3.2 Areal distribution of the Tampa and Nocatee Members, Arcadia Formation, Hawthorn Group, in southern Florida.

Lithologic Description and Areal Distribution. While the Nocatee Member was deposited at the mouth of river systems emptying into the Tampa Subsea, the sediments of the Tampa Member were deposited on the coral reefs of the Tampa Reef Tract, within the carbonate reef lagoon, and along the coral cays of the Tampa Archipelago (see Petuch, 2004: 70–77). For this reason, carbonates dominate the Tampa Member. The best lithologic description of the Tampa was given by Scott (1988: 70) and is as follows (condensed here): "... The Tampa Member consists predominantly of limestone and subordinate dolostone, sands, and clays ... Phosphate grains generally are present in the Tampa in amounts less than 3%, although beds containing greater percentages do occur, particularly near the facies change limits of the member ... the limestones are variably quartz sandy and clayey with minor or no phosphate. Fossil molds are often present and include mollusks, foraminifera, and algae. Colors range from white to yellowish-gray. The limestones range from mudstones to packstones but are most often wackestones. The dolostones often contain fossil molds, similar to those in the limestones, and are variably quartz sandy and clayey with minor or no phosphate. They are typically microcrystalline to very fine grained and range in color from pinkish-gray to light olive-gray … Siliceous beds are often present in the updip portion of the Tampa. In the type area near Tampa Bay, the unit is well known for its silicified corals, siliceous pseudomorphs of many different fossils, and chert boulders."

Siliceous pseudomorphs of gastropod mollusks have been found by divers in Tampa exposures on the stream bed of the Peace River near Arcadia, DeSoto County. The Tampa Member is geographically restricted to the west central coast of the Floridian Peninsula, from Hillsborough and Pinellas Counties, through Hardee, Manatee, Sarasota, and DeSoto Counties south to northernmost Charlotte County (Figure 3.2). Throughout its range, the Tampa varies in thickness from 82 m at Sarasota, Sarasota County, to 42 m at Bradenton, Manatee County, with an average thickness of 30.5 m. The top of the Tampa ranges in elevation from +23 m MSL in northeastern Hillsborough County to –98.5 m MSL in northern Sarasota County to over –100 m MSL in southern Manatee County.

Stratotype. From Scott (1988: 68): "… King and Wright (1979) thoroughly discussed the Tampa Member (their Tampa Formation) and its type locality. They designated Ballast Point core W-11541, Hillsborough County as the principal reference core (SE ¼, NW ¼, Sec. 11, T. 30 S., R. 18 E.). The Tampa Member occurs from –2.7 m MSL to –22.5 m MSL in this core."

Age and Correlative Units. The Tampa Member is exactly correlative with the Haywood Landing Member of the Belgrade Formation of North Carolina, the Old Church Formation of Maryland and Virginia, the lower part of the Parachucla Formation of Georgia, and the Paynes Hammock Formation of Mississippi. The Tampa Member spans the entire Aquitanian Miocene Age.

Tampa Member Index Fossils. Like those of the Suwannee Formation, the Tampa index fossils are agatized, usually being a reddish-orange or deep blue color (depending on the incorporation of either oxidized or reduced iron). The amazing richness of the Tampa fauna was summarized and illustrated by Dall (1915) and later added to by Mansfield (1937). The senior author (Petuch, 2004: 70-78) recognized and described seven distinct ecosystems preserved within the Tampa fossil fauna, including: the ***Cerion anodonta*** **Community** (Halophilic coastal grasses and salt ponds on the Tampa Archipelago); the ***Pyrazisinus campanulatus*** **Community** (Intertidal mud flats and mangrove forests); the ***Periglypta tarquinia*** **Community** (Shallow water sea grass beds); the ***Venericardia serricosta*** **Community** (Shallow sand flats); the ***Calvertitella tampae*** **Community** (Deeper water open sand sea floors); the ***Porites anguillensis*** **Community** (Sheltered back reef environments); and the ***Montastrea tampaensis*** **Community** (Main reef platform). Some of the most prominent of the Tampa index fossils are shown here on the following plates.

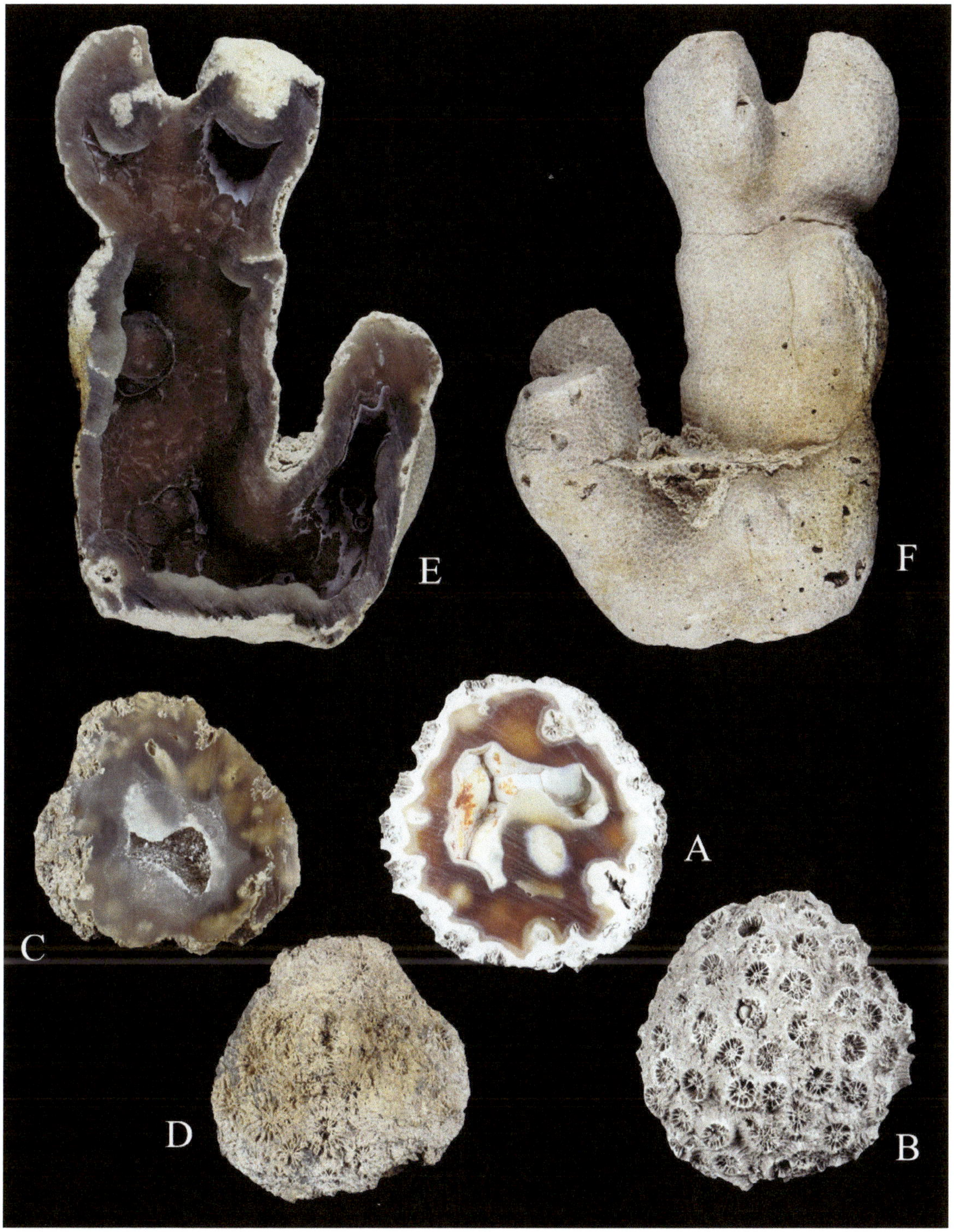

Figure 3.3 Agatized Corals of the Tampa Member, Arcadia Formation.
A, B= *Montastrea tampaensis* Vaughan, 1919, length 87 mm; **C, D**= *Antiguastrea cellulosa* (Duncan, 1863), length 71 mm; **E, F**= *Porites anguillensis* Vaughan, 1919, length 158 mm. All specimens are agatized pseudomorphs.

Figure 3.4 Mollusks of the Tampa Member, Arcadia Formation.
A, B= *Floradusta ballista* (Dall, 1915), length 23 mm; **C, D=** *Loxacypraea tumulus* (Heilprin, 1886), length 25 mm; **E=** *Chicoreus tritonopsis* (Heilprin, 1886), length 35 mm; **F=** *Vasum subcapitellum* Heilprin, 1886, length 36 mm; **G=** *Falsilyria musicina* (Heilprin, 1886), length 41 mm; **H=** *Turbo crenorugatus* Heilprin, 1886, width 28 mm; **I=** *Americoliva posti* (Dall, 1915), length 23 mm. All specimens are agatized pseudomorphs.

Figure 3.5 Mollusks of the Tampa Member, Arcadia Formation.
A= *Tritonopsis biconica* Dall, 1915, length 27 mm; **B**= *Ecphorosycon tampaensis* (Dall, 1892), length 21 mm; **C**= *Calvertitella tampae* (Heilprin, 1886), length 64 mm; **D**= *Lirophora ballista* (Dall, 1903), length 22 mm; **E**= *Anadara latidentata* (Dall, 1898), length 30 mm; **F**= *Venericardia serricosta* (Heilprin, 1886), length 28 mm; **G**= *Periglypta tarquinia* (Dall, 1900), length 44 mm; **H**= *Eucrassitella silicata* (Dall, 1915), length 34 mm; **I**= *Ampullinopsis streptostoma* (Heilprin, 1886), length 34 mm. All specimens are agatized pseudomorphs.

Figure 3.6 Mollusks of the Tampa Member, Arcadia Formation.
A= *Crassostrea vaughani* (Dall, 1915), length 47 mm; **B**= *Crassostrea rugifera* (Dall, 1898), length 52 mm; **C**= *Crassostrea rugifera pauciplicata* (Dall, 1898), length 42 mm; **D**= *Pyrazisinus campanulatus* (Heilprin, 1886), length 41 mm; **E**= *Melongena turricula* Dall, 1890, length 90 mm; **F**= *Cerion (Eostrophia) anodonta* Dall, 1890, length 28 mm; **G**= *Pterynotus posti* (Dall, 1900), length 43 mm; **H**= *Dauciconus planiceps* (Heilprin, 1886), length 49 mm; **I**= *Eichwaldiella atacta* (Dall, 1915), length 31 mm. All specimens are agatized pseudomorphs.

CHAPTER 4. The Polk, Charlotte, and Murdock Subseas (Peace River, Bayshore, and Murdock Formations)

The middle Miocene (Langhian and Serravallian Ages), the late Miocene (Tortonian and Messinian Ages), and the early Pliocene (Zanclean Age) were all times of chaotic climatic conditions, with long, mild spells separated by severe cold periods (Petuch, 2004). During the late Langhian, and also during the exceptionally severe cold time and sea level drop of the early Serravallian, the entire Everglades area was emergent. Only in the late Serravallian and early Tortonian did the Okeechobean Sea basin begin to refill, resulting in the **Polk Subsea** (named for Polk County, Florida) and, further into the late Tortonian and early Messinian, in the subsequent **Charlotte Subsea** (named for Charlotte County, Florida). After the late Messinian regression, this pattern was also repeated in the early Zanclean, by the reflooding of the Everglades area with the **Murdock Subsea** (named for Murdock, Charlotte County). The late Miocene and early Pliocene were also the times of the great deltas (Cunningham *et al.*, 2003), when the Okeechobean Sea basin was rapidly filled with the sediments of two giant deltaic complexes (discussed later in this chapter).

Throughout most of the Serravallian Miocene, during the great mid-Miocene cold time and sea level drop (Cunningham *et al.*, 2003; Petuch, 2004: 261; Petuch, 2012), the Okeechobean Sea basin remained emergent. At this time, the Everglades area was a shallow depression, probably containing one or more large freshwater lakes. In the latest Serravallian, the marine climate rapidly warmed and sea levels rose quickly (Petuch, 2012), flooding the Everglades Basin with the Polk Subsea. Existing prior to the formation of the deltas, the Polk Subsea was a deep, essentially circular lagoon bounded by a series of narrow, shallow banks (Figure 4.1). In the northwest, a series of broad, shallow lagoons, encompassed most of the coastal area of southern Florida mainland. As in the older Tampa Subsea, permanent upwelling systems occurred just offshore, flooding the main Okeechobean basin with high-productivity, plankton-rich water. Accumulating on the floor of the central lagoon, these plankton oozes became the primary source of the phosphates and phosphorite beds of the Peace River Formation.

Along the western side of the Okeechobean basin, the chain of low, shallow oyster-covered banks that had formed during the times of the Arcadia and Polk Subseas (the **Collier Banks**, discussed later in this chapter), expanded to become a larger and more prominent feature. Feeding on the rich phytoplankton resources, these oyster bioherms grew to massive proportions. Presaging the deltas that formed in subsequent subseas (discussed later in this chapter), large amounts of clay minerals were pouring into the Okeechobean basin at this time, being discharged from rivers draining into the central lagoon. These clays and muds, mixed with plankton-derived phosphates, produced the Everglades component of the Peace River Formation.

Figure 4.1 Simulated satellite image (altitude: 100 miles) of the Floridian Peninsula during the late Serravallian and early Tortonian Miocene, showing the probable appearance of the **Polk Subsea** of the Okeechobean Sea. A long, shallow area, the **Collier Banks**, had now emerged along the western side of the subsea basin.

Peace River Formation, Hawthorn Group. Although known for almost a century as a source of phosphate fertilizers (as the Hawthorn Formation or Hawthorn clays, informally named by Dall, 1892), the Peace River Formation was not formally described or defined until 1988, when Scott published his monograph on the Hawthorn Group. From this thick sequence of sediments and distinct phosphorite beds, he recognized a lower type Peace River, which makes up most of the formational sequence, and an upper Bone Valley Member, which is essentially nonexistent in the Everglades area. The Bone Valley Member, which is prevalent in northern and central Florida and is outside the scope of this book, was first recognized by Matson and Clapp in 1909, but was referred to only as the "Bone Valley Gravel." Scott's Peace River, then, is a combination of the older and poorly defined "Hawthorn Formation" and "Bone Valley Formation."

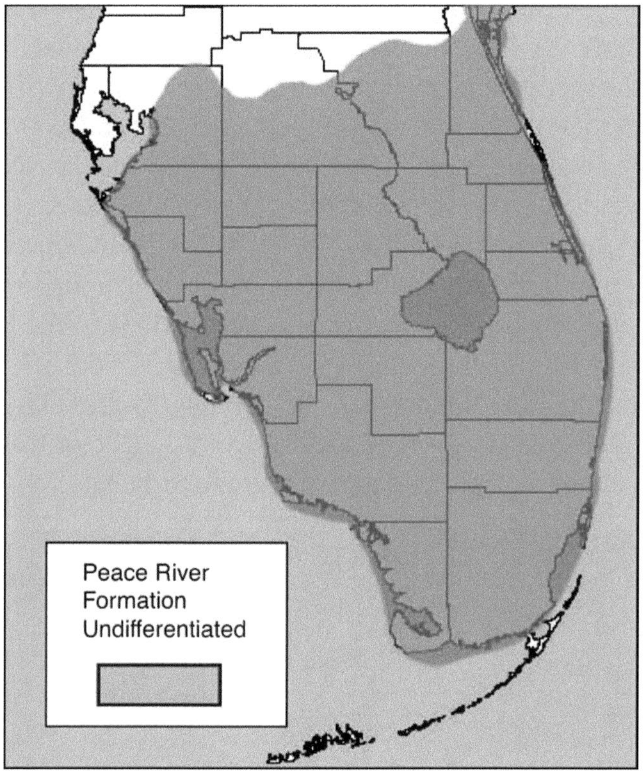

Figure 4.2 Areal distribution of the Peace River Formation (undifferentiated), Hawthorn Group.

Scott (1988: 79) also included the Bayshore and Murdock Formations of Hunter (1968, originally described as members of the Tamiami Formation) in his Peace River Formation, as the uppermost components (also referred to as the "Upper Peace River" by Cunningham *et al.*, 2003). Subsequent studies within the Everglades area, however, have shown that the areal extents, thicknesses, and developments of these formations were larger and more complicated than previously thought. As the Bayshore and Murdock are so closely tied to, and essentially define, the DeSoto and Immokalee Deltas (discussed later in this chapter), Petuch and Roberts (2007) considered them to be full, separate formations and recommended that they be removed from the Peace River (as the "Upper Peace River"). Because of the removal of the Messinian Bayshore and the Zanclean Pliocene Murdock components, the Peace River Formation now is chronologically more restricted than originally thought, dating only from the late Serravallian and early Tortonian Miocene, and not the Messinian Miocene and Zanclean Pliocene, as originally proposed. As pointed out by Scott (1988: 86), the Peace River below the Everglades region is undifferentiated, with well-defined Bone Valley sediments occurring only rarely and sporadically.

Lithologic Description of the Peace River Formation. As described by Scott (1988: 79) (condensed here): ... "The Peace River Formation consists of interbedded quartz sand, clays, and carbonates. The siliciclastic component predominates and is the distinguishing lithologic feature of the unit. Typically, the siliciclastics comprise two thirds of the formation. The quartz sands are characteristically clayey, calcareous to dolomitic, phosphatic, very fine to medium grained, and poorly consolidated. Their color ranges from light gray and yellowish-gray to olive-gray. The phosphatic content of the sands is highly variable. In the type section, the phosphate content is lowest in the upper part of the section and greatest near the base ... The phosphate occurs as sand and gravel-sized particles. The gravels are most abundant in the Bone Valley Member, although they may occur elsewhere in the unit ... Clay beds are quite common in the Peace River Formation ... the clay minerals consisting of smectite (montmorillonite), palygorskite (attapulgite), and sepiolite ... smectite and palygorskite are the dominant clay minerals in the formation ... Carbonates occur throughout the Peace River Formation. Characteristically, they comprise less than 33% of the Peace River section. The carbonates may be either limestone or dolostone. Updip (northward), dolostone occurs more frequently. The limestones are variably sandy, clayey, and phosphatic, poorly indurated, mudstones to wackestones. They vary in color from yellowish-gray to white ... Mollusk molds are common throughout the carbonates."

Tooth from the giant shark, *Otodus megalodon*, length 112 mm. Teeth from megalodon sharks, and other large elasmobranchs, are frequently found in exposed beds of the Peace River Formation.

The Peace River Formation underlies the entire Everglades region (Figure 4.2). Here, it occurs at depths of approximately 40–46 m below mean surface. The thickness of the Peace River within the Everglades Basin reflects the bathymetry of the Polk Subsea. In the west, along the shallower areas near the oyster banks, the formation has thicknesses that average 100–122 m. In the east, within the deepwater area of the Loxahatchee Trough of the Everglades (shown on Figure 6.1, Chapter 6), thicknesses may vary between 198–200 m. Unaware of the existence of this deep paleo-trough feature at the time, Scott (1988: 81) remarked that the Peace River sequence in Palm Beach County seemed "anomalously thick." Farther north, along the northern shoreline of the Polk Subsea, the Peace River occurs closer to the surface, often only at 10–20 m depths.

Stratotype. From Scott (1988: 79): ... "The type section of the Peace River Formation is designated as core W-12050, Hogan #1, located in east central DeSoto County, Florida (SE ¼, NW ¼, Sec. 14, T. 38 S., R. 23 E.), with a surface elevation of 62 feet (19 m)."

Age and Correlative Units. Based on the presence of Hemphillian mammal fossils (Scott, 1988: 84) and on Maryland Miocene-type bivalves (St. Mary's Formation, recognizable molds and casts; Petuch and Roberts: 48-49), the Peace River Formation is now known to span the period from the late Serravallian to the early Tortonian Miocene, a range of over 4 million years. The formation is exactly correlative with the St. Mary's Formation, Chesapeake Group of Maryland and with the Shoal River Formation of the Alum Bluff Group of the Florida Panhandle (Petuch, 2004; Petuch and Roberts, 2007: 48).

Peace River Index Fossils. The Peace River sediments in the Everglades area were heavily leached by infiltrating groundwater from the subsequent DeSoto and Immokalee Deltas, particularly with the development of distributaries systems and river sloughs during emergent times. Because of this fresh water infiltration and dissolution, molluscan fossils are present only as casts and molds. Some of the more recognizable moldic index fossils of the undifferentiated Peace River Formation include the mytilid *Perna incurvus* (often present in large beds), the pholadid *Cyrtopleura arcuata*, and the venerid *Clementia inoceriformis*. These species, which are common and diagnostic fossils in the Tortonian Miocene St. Mary's Formation of Maryland, are shown here on Figure 4.4. The Peace River Formation is best known for its well-preserved vertebrate fossils, especially *Otodus megalodon* shark teeth and dugong (*Metaxytherium*) ribs and teeth.

Figure 4.3 View of the bank of the Peace River, near Arcadia, DeSoto County, Florida, showing the undifferentiated Peace River Formation (the gray clay layer covered with tree roots) lying directly upon the massive limestone of the Nocatee Member of the Arcadia Formation. Photo by Rudy Wobito.

Figure 4.4 Index Fossils of the Peace River Formation.
A, B= *Akleistostoma ductor* Petuch and Drolshagen, 2011, length 52 mm; **C**= *Calusacypraea polkensis* Petuch and Drolshagen, 2011, length 42 mm; **D**= *Cyrtopleura arcuata* (Conrad, 1841), length 122 mm; **E**= *Clementia inoceriformis* (Wagner, 1839), length 68 mm; **F**= *Perna incurvus* (Conrad, 1839), length 120 mm; **G**= *Metaxytherium* Dugong rib fragment, length 140 mm. The mollusks are all steinkern casts and the bivalves were previously only known from the Miocene of Maryland and Virginia.

Deposition within the Charlotte Subsea. During the late Miocene (earliest Messinian age), a series of warm climatic intervals led to higher sea levels and a reflooding of the Okeechobean Sea basin. This new marine system, the **Charlotte Subsea**, was geomorphologically complex and represented an enlarged and infilled version of the older Polk Subsea (Figure 4.5). In the intervening emergent time, between the Polk and Charlotte Subseas (mid-to-late Tortonian sea level drop), a large deltaic complex, the **DeSoto Delta** (named for DeSoto County; Petuch and Roberts, 2007), began to feed into the Okeechobean Sea basin from a paleovalley, in what is now the DeSoto Plain and Lake Wales Ridge region (Cunningham *et al.*, 2003). This huge clastic wedge spread to fill the entire western half of the Okeechobean basin and fed into a freshwater lake that existed along the southern edge of the Everglades. Infiltrating groundwater from this long-lived lake and delta complex leached and dissolved away all the aragonitic and calcitic fossils that occurred in the underlying Peace River sediments. During the early Messinian, the DeSoto Delta was flooded by the rising sea levels of the Charlotte Subsea. This drowned delta formed a sedimentary platform upon which unique Charlotte marine ecosystems became established (Petuch, 2004: 130).

The northern edge of the Charlotte Subsea was bordered by three lagoon systems: the **Charlotte Lagoon System** (named for Charlotte County) in the west, the **Kissimmee Embayment** (named for the Kissimmee River) in the center, and the Wabasso Lagoon System (named for Wabasso, Indian River County) in the east. The western edge of the DeSoto Delta was bordered by the **Immokalee Archipelago**, a chain of low sand islands and oyster bars (bioherms) that extended from present-day Port Charlotte southward to Naples. This archipelago prevented the spreading of the deltaic sediments westward into the Gulf of Mexico, in turn, funneling this clastic wedge into the main basin of the Okeechobean Sea. The eastern half of the Charlotte Subsea basin still contained a deep water (over 200 m) lagoon, the Loxahatchee Trough, and this depression, which resembled a miniature Tongue-of-the-Ocean (Bahamas), was bordered on the east by a narrow, curving chain of oyster and coral bioherms, the Palm Beach Banks and the Miami Banks. The nearly enclosed Charlotte Subsea was connected to the Gulf of Mexico by the broad, shallow Caloosahatchee Strait (named for the Caloosahatchee River) in the west and by the deep Loxahatchee Strait (named for the Loxahatchee River) in the east. Once established in the Tortonian and Messinian, the Caloosahatchee Strait, Kissimmee Embayment, Loxahatchee Strait, and Loxahatchee Trough continued to persist in all the subsequent subseas until the end of the Pleistocene. Of special interest during this time was the existence of extensive and long-lived upwelling systems along the western coast of Florida. These cool and nutrient-rich water masses, which were probably produced by a strong, southward-flowing branch of the **Apalachee Gyre** (Petuch, 2004: 157), supported immense phytoplankton blooms all along the Charlotte Lagoon System and the Immokalee Archipelago. These high-productivity waters and their plankton biomass were the principal sources for the phosphorites and phosphate grains that are so prominent in the sediments of the Bayshore Formation.

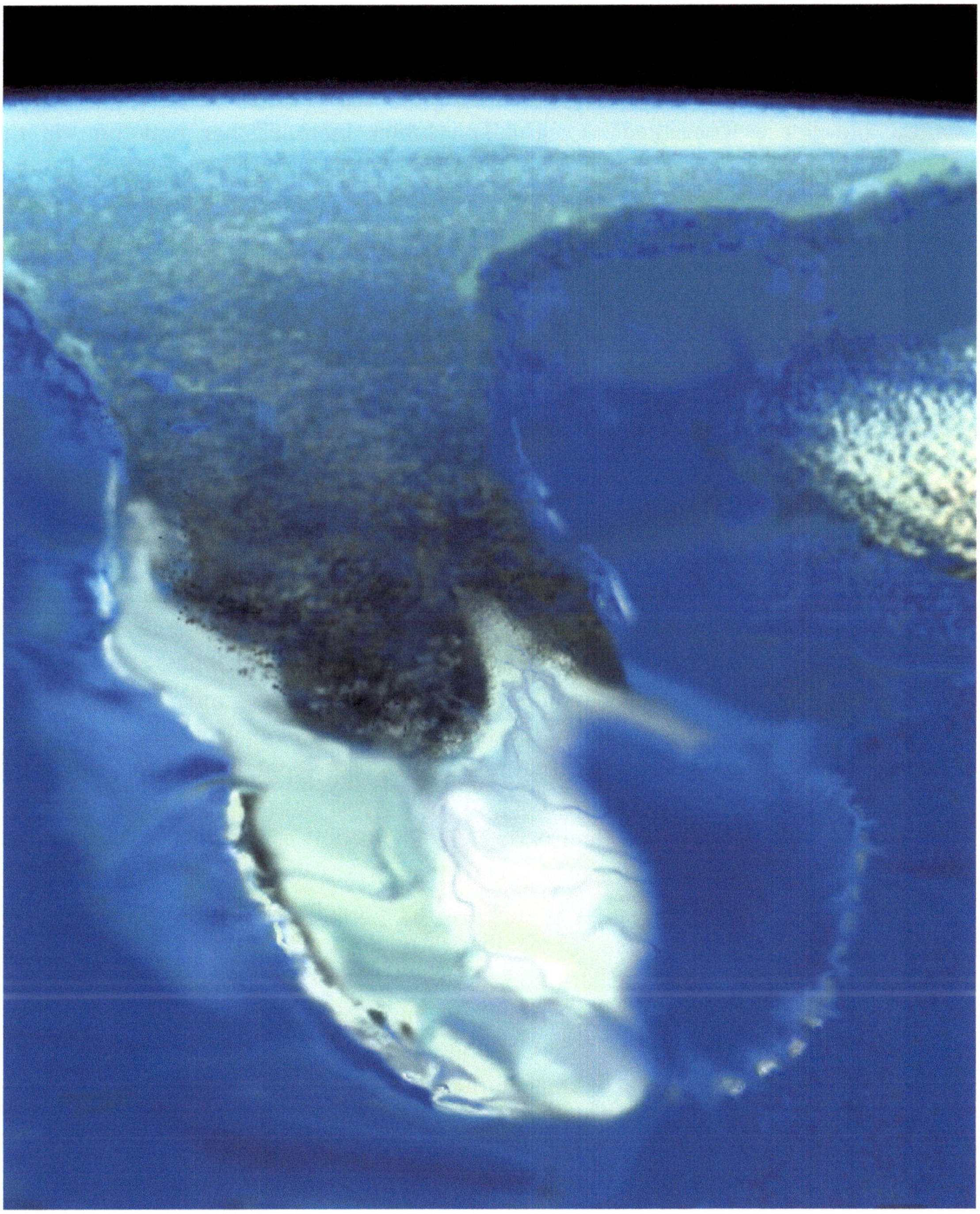

Figure 4.5 Simulated satellite image (altitude: 100 miles) of the Floridian Peninsula during the Messinian Miocene, showing the possible appearance of the **Charlotte Subsea** of the Okeechobean Sea. At this time, the large **DeSoto Delta** had developed along the southwestern end of the peninsula and extended across the western half of the Charlotte Subsea. Along the western side of the delta, the old Collier Banks had now developed into the **Immokalee Archipelago**, a long chain of oyster banks and low islands. To the north of this area, a large estuary, the **Charlotte Lagoon System**, was developing.

Bayshore Formation, Hawthorn Group. Until recognized as a set of distinctive, phosphate-bearing beds by Hunter in 1968, this formation was generally overlooked by most previous workers or simply considered to be part of the older Miocene Bone Valley and Peace River beds. Originally, Hunter included her new unit in the Tamiami Formation, Okeechobee Group, considering it to be the lowest and oldest member. As discussed previously, Scott (1988) later placed Hunter's Bayshore "Member" and Murdock Station "Member" (discussed later in this chapter) in his Peace River Formation, Hawthorn Group (as the "Upper Peace River"). Because of the presence of well-preserved mollusks and barnacles, which constitute a large carbonate fraction, the Bayshore can be differentiated from the underlying Peace River and can be seen to constitute a separate geologic and lithologic unit. The formation was recently removed from member status in the carbonate-rich Tamiami Formation, Okeechobee Group (as proposed by Hunter), and was elevated to full formational status in the phosphate-rich Hawthorn Group (Petuch, 2004: 18).

Lithologic Description of the Bayshore Formation. Overall, the Bayshore Formation is lithologically distinctive and is easily separable from the underlying Peace River Formation and the overlying Murdock Formation. Typically, the Bayshore is composed of clays (smectite and palygorskite) and sandy clays, usually white, tan, or pale orange in color, and contains variable amounts of black phosphorite grains and phosphate-coated pebbles. Some facies, such as those encountered in canal digs near Murdock, Charlotte County, are sandier, light gray in color, and contain higher amounts of phosphates. Characteristically, thick beds of well-preserved mollusk shells (typically scallops, oysters, and ecphoras) and barnacles (Cirripedia) are present and make up a large percent of the carbonate fraction. The formation was found to be better developed and more widely distributed than was originally thought by Hunter, and has been shown to consist of two separate members; the lower Cocoplum Member (early Messinian) and the upper Port Charlotte Member (late Messinian), which are described in the following sections. The Cocoplum Member, which is separated by a large unconformity from the overlying Port Charlotte Member, also contains sandy dolomite beds that contain large amounts of phosphorites.

These beds vary in color from light orange and buff to pale gray. (Hunter, original description: "… white to light tan in color; sandy to very sandy clay containing black and brown phosphate grains and small pebbles. Well-preserved calcite fossils are present …"). The Bayshore Formation, including both the Cocoplum and Port Charlotte Members, underlies most of the Everglades region (Figure 4.6). The thickest and best-developed areas of the Bayshore are known primarily from the Port Charlotte and southern Sarasota County areas. Here, the formation is estimated to be over 2 m thick and in some areas, such as along the Cocoplum Canal, southern Sarasota County, the formation may reach thicknesses of over 3 m. In all these localities, the Bayshore occurs at depths ranging from 4 to 7 m below mean surface. Based on the distribution of key molluscan index fossils, the formation is now known to extend from Pinellas County

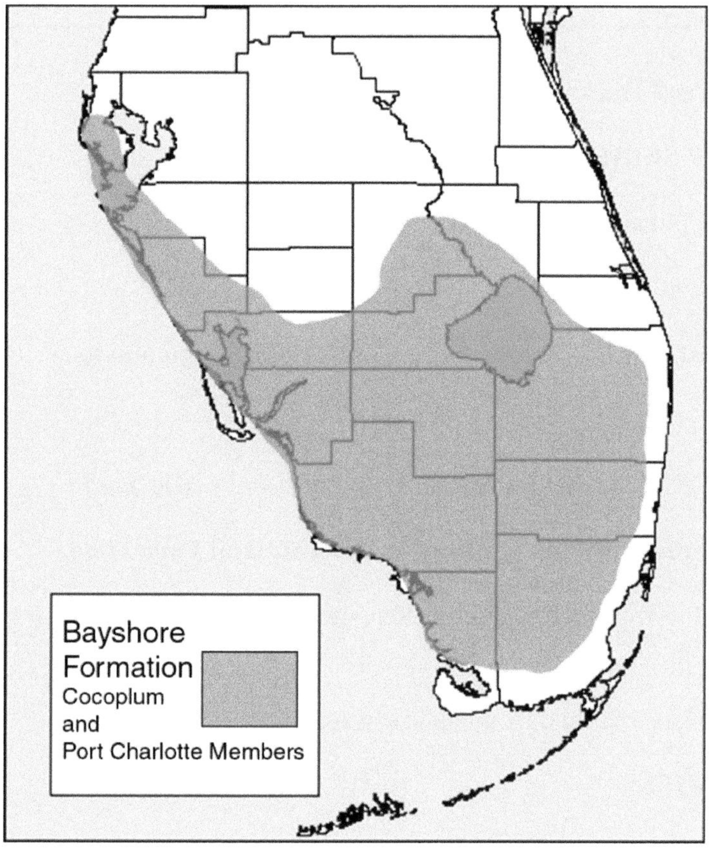

Figure 4.6 Areal Distribution of the Bayshore Formation, Hawthorn Group, and its two members, the Cocoplum and the Port Charlotte.

south to northern Monroe County, and from Okeechobee and Highlands Counties in the Kissimmee Valley south to Dade County. Within the Okeechobee Plain areas, Bayshore index fossils have been encountered in deep dredging (approximately 20 m) by draglines during Kissimmee River channel deepening. The smectite and palygorskite (brackish water) clays of the Bayshore Formation are distinctly deltaic and reflect the depositional influence of the DeSoto Delta throughout the Messinian. Subsequent groundwater infiltration from the Immokalee Delta during the Zanclean Pliocene led to further leaching and diagenesis. This deltaic leaching also dissolved all Bayshore aragonitic fossils, leaving only calcitic species.

Stratotype. From Hunter (1968: 443): "... Canal bank, Section 3, T. 40 S., R. 21 E., on Bayshore Waterway, ½ mile south of U.S. 41, Port Charlotte, Charlotte County, Florida." The formation was described from an underwater exposure, from near the surface to approximately 1 m below surface. A classic stratigraphic column for the Bayshore sequence was given by Hunter (1968: 443) and is reproduced here as follows (with modifications by ourselves in light of new data):

Bed Description and Thickness (in feet)

PLEISTOCENE SERIES

Fort Thompson Formation

Bed 6 Sand 4.0 ft

Bed 5 Shelly sand with *Pyrazisinus gravesae* (originally referred to as "*P. scalatus*") 2.0 ft

Bed 4 Sand 1.5 ft

EARLY PLIOCENE SERIES (originally thought to be Miocene)

Murdock Formation (originally placed in the Tamiami Formation)

Bed 3 Pecten biostrome, dark gray pecten shells and a few barnacle fragments (lower bed of Murdock) 4.5 ft

Bed 2 Gray phosphatic sand to and below water level 2.5+ ft

MIOCENE SERIES

Bayshore Formation (Port Charlotte Member)

Bed 1 Sandy clay from below water level containing *Globecphora floridana* (originally referred to as "*Ecphora quadricostata*") and *Chesapecten middlesexensis hunterae* (originally referred to as "*Pecten santamaria middlesexensis*") 1.0+ ft

---

Age and Correlative Units. Based upon the presence of key index fossils, such as *Ecphora whiteoakensis hopei*, *Chesapecten middlesexensis hunterae*, and *Carolinapecten urbannaensis*, the Bayshore Formation is now known to date from the latest Miocene (Messinian). The Bayshore is exactly correlative with the Eastover Formation, Chesapeake Group of Virginia and northern North Carolina. The two members of the Bayshore Formation, the Cocoplum and Port Charlotte, are separated by an unconformity that represents almost 1 million years of missing time.

Cocoplum Member, Bayshore Formation.

Lithologic Description and Areal Distribution. The Cocoplum Member characteristically contains large amounts of sandy dolosilts and thick sandy dolomite stringers, most often of an orange or a tan-buff color, but sometimes white, and containing variable amounts of black phosphorite granules. Clays, most commonly palygorskite and smectite, are also present in variable amounts and are colored a light tan or orange-tan. Molluscan bioclasts are abundant in this member, consisting mostly of oysters and muricid gastropods, and make up the largest part of the carbonate fraction. Gastropod fossils from this member are colored a distinctive golden yellow or pale orange. The uppermost part of the Cocoplum Member indurates into a semifriable, grayish sandy dolomite/arenaceous calcarenite that is rich in phosphate grains. Oysters from this indurated unit (Figure 4.7) are usually colored a gray-tan. The Cocoplum Member has the same areal distribution as the entire Bayshore Formation, extending from Pinellas County to Monroe and Dade Counties. The classic Cocoplum index fossils *Mansfieldostrea brucei* and *Ecphora whiteoakensis* (subspecies *hopei*) have been collected during dredgings in the Kissimmee River (20 m depth) near the Osceola–Okeechobee County line. Throughout its whole range, the Cocoplum Member has thicknesses that average around 1.5 m but may be thicker in the northern Kissimmee River Valley.

Stratotype. The type section of the Cocoplum Member is found along the eastern side of the Cocoplum Waterway Bridge at North Port, Sarasota County, on the Cocoplum Waterway just north of the Sarasota County–Charlotte County line. The type sedimentary material and index fossils are exposed at times of lowest water level, just above the water surface (Hunter, 1968: 443). The new member is named for the Cocoplum Waterway, Sarasota County, the type locality. Cocoplum sediments and index fossils were collected along 2 km of the canal bank.

Age and Correlative Units. The Cocoplum Member is now known to date from the early Messinian Miocene and is exactly correlative with the Claremont Manor Member of the Eastover Formation, Chesapeake Group of Virginia and northern North Carolina. Both the Claremont Manor and the Cocoplum Members share the chronologically sharply-defined bivalves *Mansfieldostrea brucei*.

Cocoplum Index Fossils. As the Cocoplum sediments are highly leached, only thick-shelled calcitic fossils are preserved. A single Cocoplum marine community was described by the senior author (Petuch, 2004: 132), the ***Gigantostrea leeana* Community** (shallow sand-bottom lagoon and oyster bars). Some of the principal Cocoplum index fossils are shown here on the following figure. A particularly interesting and important event occurred during Cocoplum time, when entire shoals of large multi-armed *Heliaster* sun sea stars appeared to have died simultaneously, forming an enormous death assemblage (thanatocoenosis). This still-undescribed species of sun star apparently lived in abundance on the extensive oyster and barnacle bioherms that occurred right offshore and succumbed, *en masse*, to some type of catastrophic event, possibly a hurricane or extreme cold weather event. The dead and dying sun stars washed into shallow water, or onto the beach, and accumulated in thick beds composed of hundreds of thousands of individuals. After the dead sun stars packed together into a solid layer, their calcitic skeletal ossicles fused together to form a unique type of echinoderm limestone that is only known from the Cocoplum Member. A fragment of this *Heliaster* echinoderm limestone is shown here on Figure 4.7.

Figure 4.7 Mollusks and Echinoderms of the Cocoplum Member, Bayshore Formation.
A= *Ecphora whiteoakensis hopei* Petuch, 1991, length 101 mm; **B=** *Gigantostrea leeana* (Wilson, 1987), length 138 mm; **C=** *Zulloia zulloi* Petuch, 1994, length 36 mm; **D=** *Zulloia violetae* Petuch, 1994, length 34 mm; **E=** *Zulloia* unnamed species with winged varices, length 33 mm; **F=** *Mansfieldostrea brucei* Ward, 1992, length 97 mm; **G=** Sun Star *Heliaster* unnamed species, ventral view showing mouth and ambulacral grooves on the base of the arms, length 162 mm.

Port Charlotte Member, Bayshore Formation

Lithologic Description and Areal Distribution. The Port Charlotte Member characteristically is composed of sandy clay (containing smectite, palygorskite, and sepiolite) that is colored white, yellow, or tan, with large amounts of black phosphate grains and phosphate-coated pebbles. Some facies contain large amounts of highly phosphatic quartz sand, with very little clay, and are light gray or tannish-gray in color. Typically, the Port Charlotte contains abundant well-preserved molluscan bioclasts, predominantly large scallops, oysters, and muricid gastropods, which make up the greatest part of the carbonate fraction. Dolosilts are occasionally present, but make up only a minor fraction of some facies. The Port Charlotte Member has the same areal distribution as the entire formation, extending from Pinellas County to Monroe and Dade Counties and into the Kissimmee Valley of the Okeechobee Plain. Throughout its range, the Port Charlotte has thicknesses that average around 1–2 m (Petuch and Roberts, 2007: 54).

Stratotype. The type section of the Port Charlotte Member is the canal bank on Bayshore Drive, along the Bayshore Canal, Port Charlotte, Charlotte County, Florida (Sec. 3, T. 40 S., R. 21 E., Charlotte County). Here, the Port Charlotte Member is exposed at water level (during low-water times) in the Bayshore Canal, approximately 4 m below land surface, and has a mean thickness of 1.5–2 m. The new member was named for Port Charlotte, Charlotte County, Florida (Petuch and Roberts, 2007: 56).

Age and Correlative Units. The Port Charlotte Member is now known to date from the late Messinian Miocene and is exactly correlative with the Cobham Bay Member of the Eastover Formation, Chesapeake Group of Virginia and northern North Carolina. Both the Port Charlotte and Cobham Bay Members share the chronologically sharply-defined mollusks *Carolinapecten urbannaensis*, and *Mansfieldostrea geraldjohnsoni*, all classic guide fossils for the late Messinian Miocene of the Atlantic Coastal Plain.

Port Charlotte Index Fossils. Like the Cocoplum Member, the Port Charlotte is highly leached and contains only thick, resistant calcitic fossils. A single marine community from the Port Charlotte Member was described by the senior author (Petuch, 2004: 132), the *Chesapecten middlesexensis* Community (deep sand-bottom lagoon scallop beds); now emended to *Chesapecten middlesexensis hunterae*; Waller, 2018. The principal index fossils for the Port Charlotte member are shown on the following figure.

Figure 4.8 Mollusks of the Port Charlotte Member, Bayshore Formation.
A= *Calusathais handgenae* (Portell and Vokes, 1992), length 58 mm; **B**= *Globecphora floridana* (Petuch, 1989), length 110 mm; **C**= *Globecphora floridana streami* (Petuch, 1994), length 112 mm (confined to the uppermost bed of the Port Charlotte Member); **D**= *Mansfieldostrea geraldjohnsoni* Ward, 1992, length 94 mm; **E**= *Chesapecten middlesexensis hunterae* Waller, 2018, length 120 mm; **F**= *Carolinapecten urbannaensis* (Mansfield, 1929), length 145 mm.

Deposition within the Murdock Subsea. The latest Messinian Miocene was a time of rapidly-repeating severe cold times and major sea level drops, and this led to the extinction of many classic Bayshore molluscan genera, such as *Zulloia*, *Calusathais*, and *Gigantostrea*. Subsequently, the Okeechobean Sea basin remained emergent for over 1 million years, from approximately 5.5 million to 4.5 million years B.P. and marine climates did not become warmer until the early Zanclean Pliocene (5.3 million years). At that time, sea levels again began to rise, and southern Florida was reflooded as the **Murdock Subsea** (Figure 4.9). During the intervening emergent time in the latest Messinian, the southwestern part of the Okeechobean basin was covered by a second large deltaic system, the **Immokalee Delta** (named for Immokalee, Collier County; see Petuch and Roberts, 2007). This prominent geomorphological feature prograded over the older late Miocene DeSoto Delta, filling all deeper depressions along the western side of the Everglades with terriginous sediments. During the reflooding of the Okeechobean basin in the Zanclean Pliocene, the Immokalee Delta was drowned and produced a shallow platform that encompassed the western two thirds of the Murdock Subsea.

Several persistent late-Miocene geomorphological features became larger and more prominent during the early Zanclean sea level rise. The oyster banks along the western Okeechobean Sea became wider at this time, expanding eastward across the platform formed by the Immokalee Delta. The narrow oyster banks along Palm Beach and Dade Counties also began to shoal at this time, producing large, shallow-water oyster and coral bioherms. These enlarged Charlotte Subsea features, now form the core of the southern section of the present-day Atlantic Coastal Ridge of southeastern Florida. As in the older Charlotte Subsea, large upwellings of cold, nutrient-rich water occurred along the western side of the Murdock Subsea. Besides producing colder water conditions, these upwellings flooded the shallow northern section of the Murdock Subsea with plankton-rich water, leading to the production of phosphorites. These high-productivity waters supported rich and highly endemic marine communities, comprising immense beds of giant *Chesapecten* scallops, extensive giant barnacle (Cirripedia) "reefs," massive oyster bars, and a rich fauna of marine mammals, including walruses, seals, and whales. The principal depositional unit of the Murdock Subsea, the Murdock Formation, represents the last vestige of phosphate deposition and marks the top of the Hawthorn Group.

Murdock Formation, Hawthorn Group. Like the underlying Bayshore Formation, this last phosphate-bearing set of beds was generally overlooked by previous authors until recognized and described by Hunter (1968). Despite the presence of phosphorites, Hunter included this poorly studied unit within her expanded "Tamiami Formation". Originally considered to be the second-lowest member of the Tamiami Formation, Hunter named this unit the "Murdock Station Member," with the stratotype near Murdock, Charlotte County (Murdock Station was the old railroad station for Port Charlotte). Like the Bayshore Formation discussed previously, the Murdock Formatio was placed within the Peace River Formation by Scott (1988) and

Figure 4.9 Simulated satellite image (altitude 100 miles) of the Floridian Peninsula during the Zanclean Pliocene, showing the probable appearance of the **Murdock Subsea** of the Okeechobean Sea. At this time, a larger deltaic system, the **Immokalee Delta**, developed on top of the old DeSoto Delta and filled almost three-quarters of the Okeechobean Sea Basin.

later considered part of the "Upper Peace River" (Cunningham *et al.*, 2003). Because of its unique lithology and larger development than what was originally thought, this formation was given full formational status by the senior author (Petuch, 2004: 18) and by Petuch and Roberts (2007: 59). Since the original name, "Murdock Station Formation", is considered cumbersome, we here emend the name to the "Murdock Formation", which reflects the original intent of Hunter for naming it after the town of Murdock.

Lithologic Description and Areal Distribution. This uppermost component of the Hawthorn Group is lithologically distinct across its entire extent and is easily separable from the overlying Tamiami Formation and the underlying Bayshore Formation. As described by Hunter (1968), the type Murdock is composed of coarse beach-type quartz sand mixed with small amounts of clays (primarily smectite, palygorskite, and sepiolite), black phosphate grains, and lime muds (calcilutites), and is generally a dark gray or gray-brown color. Subsequent research has shown that this formation comprises two separate members: the lower **Jupiter Waterway Member** (early Zanclean) and the upper **Sarasota Member** (late Zanclean) (see Petuch and Roberts, 2007).

The Murdock Formation locally indurates into a dark gray phosphatic arenaceous limestone containing molds of bivalve and gastropod mollusks. At the Quality Aggregates quarries at Sarasota, the upper surface of the Sarasota Member (the top 0.5 m, in contact with the base of the overlying Buckingham Member of the Tamiami Formation), is indurated into a dark khaki-gray moldic arenaceous limestone. The Murdock Station differs from the Peace River Formation in having extremely thick, well-developed, and well-preserved beds of mollusks, barnacles, and ahermatypic corals, and these make up the majority of the carbonate fraction. Both members of the underlying Bayshore Formation differ in having thinner, less-developed shell beds (and a corresponding lower carbonate fraction) and containing a higher percentage of clay minerals. As pointed out by Hunter (1968: 443), two distinct sets of beds are present in the Murdock: an upper set containing gray oyster–barnacle hash, fine to medium black phosphate grains, and clayey quartz sand, and a lower set, containing dense beds of large *Chesapecten* scallops, gray to brown clayey quartz sand with abundant phosphorite grains and phosphate-coated pebbles, and stringers of highly phosphatic unfossiliferous gray sand. Ketcher (1992), in her study of Unit 11 at Sarasota (see Chapter 5), also recognized a lower scallop-rich zone and an upper barnacle-rich zone, but added an uppermost shell bed composed of *Diplodonta* clams. At Quality Aggregates pit #6 at Sarasota, the upper set of beds contained massive "reefs" (bioherms) of the giant barnacle *Chesaconcavus* and intertwined large bioherms of the branching ahermatypic coral *Oculina sarasotana*. The lower set of beds ("Unit 12"; see Chapter 5) contained dense aggregations of the giant scallops *Chesapecten jeffersonius* and *Chesapecten quinarius*, and the gryphaeid oyster *Mansfieldostrea compressirostra*.

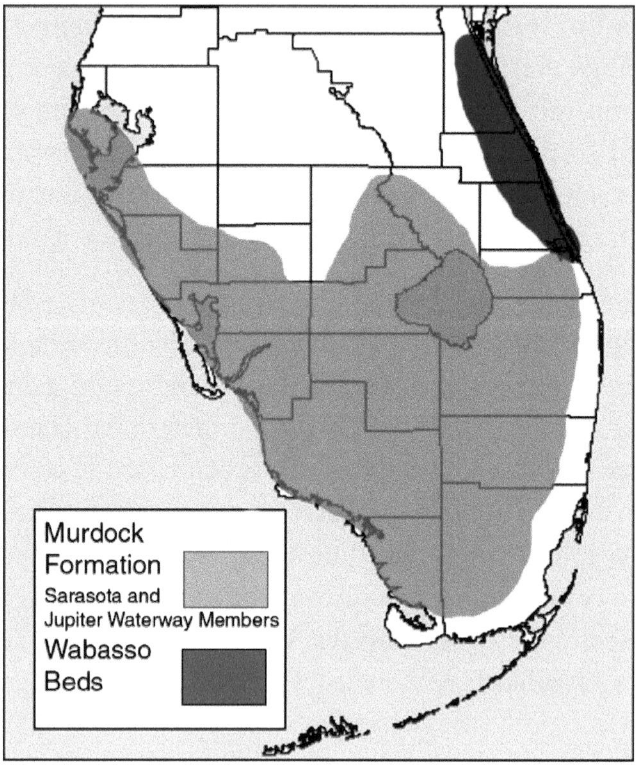

Figure 4.10 Areal distribution of the Murdock Formation, Hawthorn Group and its two members, the Jupiter Waterway and the Sarasota, in southern Florida. The "Wabasso Beds" facies can be seen to extend up the eastern coast.

The Murdock Formation appears to have a wider areal extent than does the underlying Bayshore Formation. Both members of the Murdock have the same distribution, and extend from Pinellas County in the north to Monroe and Dade Counties in the south, and throughout the Everglades area. At the Florida Rock Industries Naples Quarry (old "Mule Pen Quarry"), East Naples, Collier County, the Murdock Station was dredged from 20 m depth and contained large numbers of the oyster *Mansfieldostrea compressirostra* and small barnacles, all in a light gray phosphatic clay. The formation extends eastward into the Caloosahatchee and Kissimmee River valleys and has been identified in shallow (5–10 m depth) dredgings at the headwaters of the Kissimmee River, near old Fort Kissimmee, Highlands County (during Kissimmee wetland remediation excavations) and on spoil piles along the Kissimmee Prairie, Okeechobee County. Here, a gray phosphatic clay containing *Chesapecten quinarius* was exposed all along the canal dig and on spoil islands in the marshland. The phosphatic "Wabasso Beds" (Scott, 1988) encountered along coastal southeastern Florida (Figure 4.10), particularly in Indian River and St. Lucie Counties, appear to belong to the Murdock Formation and represent the extreme eastern formational development. Dated as Zanclean Pliocene (Scott, 1988: 84), the Wabasso Beds may be correlative with Hunter's "Upper Bed" of the Murdock (here referred to as the Sarasota Member). Throughout its lateral extent, the Murdock Formation is variable both in thickness and in depth below mean surface. At

Sarasota, the upper formational surface (Unit 11, Figure 4.3) occurs at approximately 20 m below surface, and this same mean depth is seen at the "Mule Pen Quarry" at East Naples. At the stratotype near Murdock, in the Port Charlotte area, the upper surface of the formation occurs at much shallower depths, from approximately 2.5 m to 3 m, and averages 2 m in thickness. The formation (both members) is over 8 m thick at Sarasota but may reach thicknesses of over 10 m at Naples. The Kissimmee Murdock Station beds are closer to mean surface and may be only 1 m thick.

Stratotype. From Hunter (1968: 443): "… Canal bank, Sec. 4, T. 40 S., R. 21 E., on Jupiter Waterway, 1 mile south of U.S. Route 41, Port Charlotte, Charlotte County, Florida." The formation was described from the banks of the canal near the old railroad station at Murdock, where it was exposed 2 m above the water surface. A classic stratigraphic column for the Murdock Formation sequence was given by Hunter (1968: 443), and is reproduced as follows (with modifications by the authors in light of new data):

Bed Description and Thickness (in feet)

PLEISTOCENE SERIES

Fort Thompson Formation

Bed 8 Tan sand 1.0 ft

Bed 7 Red-brown sand 2.0 ft

Bed 6 Shelly sand containing *Pyrazisinus gravesae* (originally "*P. scalatus*") 3.0 ft

Bed 5 Brown sand 1.0 ft

PLIOCENE SERIES (Originally Referred to as the Miocene Series)

Murdock Formation (originally placed in the Tamiami Formation as a member)

Bed 4 Hard fragmental limestone with barnacle fragments (upper bed of Murdock) 3.0 ft

Bed 3 Gray phosphatic sand 0.5 ft

Bed 2 Pecten biostrome (containing *Chesapecten jeffersonius*) 0.5 ft

MIOCENE SERIES

Bayshore Formation (Port Charlotte Member)

Bed 1 Gray phosphatic sand 3.0 ft

Age and Correlative Units. Based on the presence of key chronostratigraphic index fossils such as *Chesapecten jeffersonius* and *Mansfieldostrea compressirostra*, the

Murdock Formation is now known to date from the Zanclean Pliocene. This uppermost formation of the Hawthorn Group is exactly correlative with the Sunken Meadow Member of the Yorktown Formation, Chesapeake Group ('Yorktown Zone 1") of Virginia and northern North Carolina, the lower beds of the Duplin Formation of southern North Carolina and northern South Carolina, and the lower beds of the Goose Creek Formation of southern South Carolina.

Jupiter Waterway Member, Murdock Formation. At the type section along the Jupiter Waterway, the lower bed of the Murdock Formation (Bed 2) stands out as distinctive, both lithologically and paleontologically. The high percentage of sepiolite clays and large quartzite pebbles and the presence of the giant scallop *Chesapecten jeffersonius* demonstrates that this unit constitutes a separate member of the Murdock Formation. Based on the name of the stratotype locality, Petuch and Roberts (2007; 62) proposed the name "Jupiter Waterway Member" for this conspicuous lower bed.

Lithologic Description and Areal Distribution. (from Petuch and Roberts, 2007: 62) The Jupiter Waterway Member "...characteristically is composed of gray or brownish-gray clay (mostly sepiolite) with varying amounts of gray quartz sand and containing a high percentage of black phosphorite grains and phosphate-coated pebbles (3 to 5%; see Ketcher, 1992: 170). A high carbonate fraction is present in some facies, and this is almost always composed of scallop shell and oyster bioclasts. In some areas, such as near Port Charlotte, the classic clay facies interfingers with thick lenses of unfossiliferous phosphatic gray quartz sand ("Lower Bed"; Hunter, 1968: 443). In the APAC and Quality Aggregates quarries at Sarasota (Ketcher, 1992), the basal beds of the Jupiter Waterway Member are composed of heavily bioturbated unfossiliferous clays. The phosphatic "Alva Clay" and "LaBelle Clay" of Puri and Vernon (1964) appear to be facies of the Jupiter Waterway Member. The upper bed ("Lower Zone" of Unit 11; see Ketcher, 1992) typically contains a thick, densely packed bed of large scallops, mostly *Chesapecten* species, and the oyster *Mansfieldostrea compressirostra* (the "*Pecten* Biostrome" of Hunter, 1968).

The areal distribution of the Jupiter Waterway Member is the same as that of the entire formation, extending from Pinellas County south to Monroe and Dade Counties and underlying the entire Everglades. Throughout its range, the Jupiter Waterway can be easily traced by the presence of Hunter's "*Pecten* Biostrome" in canal digs, quarries, and driller's cores. Specimens of *Chesapecten jeffersonius*, and *Mansfieldostrea compressirostra* were also found in deep well cores (70–80 m, in the area of the deep Loxahatchee Trough) from western Miami, Dade County (along U.S. 41), demonstrating that the Jupiter Waterway underlies the entire Everglades area. The new member varies in thickness from 1 m at the stratotype near Port Charlotte, Charlotte County, to 3 m at the "Mule Pen Quarry" at East Naples, Collier County, to over 10 m west of Miami, Dade County. A discernable unconformity is present at the top of the Jupiter Waterway Member, usually preserved as a thin sand or sandy limestone layer.

Stratotype. From Hunter (1968: 443): " ... Canal bank, Sec. 4, T. 40 S., R. 21 E., on Jupiter Waterway, 1 mile south of U.S. Route 41, Port Charlotte, Charlotte County, Florida." The stratotype of the Jupiter Waterway Member comprises Unit 2 ("Pecten Biostrome," type lower bed of the Murdock; Hunter, 1968) and Unit 1 ("gray phosphatic sand"; Hunter, 1968) of the stratigraphic column described in the previous section (the stratotype for the Murdock Formation).

Age and Correlative Units. Based on the presence of an entire assemblage of geochronologically restricted index fossils, the Jupiter Waterway Member is now known to date from the early Zanclean Pliocene. Some of these key early Zanclean markers include *Chesapecten jeffersonius*, *Ecphora quadricostata*, and *Mansfieldostrea compressirostra*. This phosphate-bearing unit is exactly correlative with the lower beds of the Sunken Meadow Member of the Yorktown Formation, Chesapeake Group (lower "Yorktown Zone 1") of Virginia and northern North Carolina, the basal beds of the Duplin Formation of southern North Carolina and northern South Carolina, and the basal beds of the Goose Creek Formation of southern South Carolina.

Jupiter Waterway Index Fossils. As the Jupiter Waterway sediments are heavily leached, only thick calcitic fossils are preserved. Some of the principal index species are shown here on Figure 4.11. The tusks of the Florida Walrus, *Ontocetus emmonsi*, are also frequently encountered within the beds of giant *Chesapecten* scallops, their apparent food preference.

Sarasota Member, Murdock Formation. The upper bed of the Murdock type section along Jupiter Waterway (Bed 4) is lithologically distinct, with low levels of sepiolite clays and with no fossil scallops. Instead, abundant fragments of large cirriped barnacles (*Chesaconcavus* species) make up most of the carbonate fraction, showing great differences in lithology and paleontology between this younger, upper unit and the subjacent Jupiter Waterway Member. Because the stratotype locality occurs within the APAC quarry at Sarasota, Petuch (1988) and Petuch and Roberts (2007: 63-65) proposed the name "Sarasota Member" for this upper part of the Murdock Formation.

Lithologic Description and Areal Distribution. From Petuch (1988: 120–121; originally described as the uppermost member of the "Hawthorn Formation"): "... Typically a yellow-tan calcareous clay, containing a large percentage of phosphorite pebbles and granules, and abundant phosphatized bone fragments. Bone fragments are mostly from large marine mammals such as cetaceans and sirenians. Shark teeth are also common. Exposed surfaces weather to a gray or gray-tan color. In some sections, the upper half meter is indurated into a hard, gray, phosphatic limestone, often highly fossiliferous." Subsequent research at larger and better exposures (such as at Quality Aggregates pit #6 at Sarasota) has shown that the Sarasota Member is lithologically more diverse than originally thought. The formal lithologic description is here expanded to include deposits of sandy clay (mostly smectite and sepiolite) with

Figure 4.11 Index Fossils of the Jupiter Waterway Member, Murdock Formation.
A= *Ecphora quadricostata* (Say, 1824), length 60 mm; **B**= *Chesapecten jeffersonius* (Say, 1824), length 142 mm; **C**= *Chesapecten quinarius* Waller, 2018, length 83 mm; **D**= *Carolinapecten corpulentus* Waller, 2018, length 65 mm; **E**= *Mansfieldostrea compressirostra* (Say, 1824), length 99 mm (black phosphorite grains are cemented to exterior); **F**= *Pterorhytis* species, length 45 mm; **G**= *Ecphora pachycostata* Petuch, 1989, length 76 mm; **H**= *Chesapecten middlesexensis bayshorensis* Waller, 2018, length 102 mm.

stringers of arenaceous calcilutites (sandy lime muds) and abundant black phosphate grains (3–5% phosphate; Ketcher, 1992). The largest part of the carbonate fraction is composed of bioherms of giant barnacles (Cirripedia) and densely intertwined beds of the branching ahermatypic coral *Oculina* ("coral thickets" of Ketcher, 1992: 174). The calcilutite stringers often contain well-preserved specimens of aragonitic fossils, primarily small bivalves such as *Mulinia*, *Diplodonta*, and *Pleiorhytis*. Abundant marine mammal bones are typically present, usually from whales, walruses, seals, and sirenians (dugongs and manatees).

The Sarasota Member has essentially the same areal distribution as the underlying Jupiter Waterway Member. Over its range, the Sarasota can always be differentiated from the basal Murdock beds, and also from the Bayshore Formation, by its higher percentage of carbonates (both lime muds and bioclasts) and by its lower percentage of clay minerals. The Sarasota Member reaches thicknesses of over 5 m at the "Mule Pen Quarry" at East Naples, but only 2 m at the stratotype in the APAC pit at Sarasota (Figure 4.3, Chapter 4). The upper part of the "Wabasso Beds" of Indian River and St. Lucie Counties are probably correlative with the Sarasota. The new member is exposed at the stratotype of the Murdock Formation along the Jupiter Waterway. Here, it was designated Unit 4 ("Type Upper Bed") by Hunter (1968: 443) and was found to be only 1 m in thickness.

At the type section of the Sarasota Member in Quality Aggregates pit #6, a conspicuous thin layer of gray phosphatic sandstone, approximately 10 to 20 cm in thickness, lies directly on top of the bedding plane surface of Unit 11. This small sandstone unit represents an indurated duricrust that formed after the deposition of the underlying Sarasota Member and before the deposition of the overlying Buckingham Member of the Tamiami Formation (discussed in the next chapter). Designated as "Unit 11A" (Petuch and Roberts, 2007), this phosphatic sandstone unit formed during the subaerial conditions produced by the low sea levels at the end of the Zanclean Pliocene (see Figure 5.2 in the next Chapter). A number of well-preserved aragonitic fossils have been collected from Unit 11A, including gastropods such as an undescribed *Tropochasca*, small olivids, and the small scaphelline volute, *Scaphella kendrae*. Some of these are shown here on Figure 4.12.

Stratotype. From Petuch (1988: 121): "… The type is the basal bed (Unit 11 of Petuch, 1982) of the APAC pit mine (formerly the Macasphalt and Warren Brothers pit), Newburn Road, Sarasota, Sarasota County, Florida. In the APAC pit, only one meter or less of the Sarasota Member is exposed at 20 m depth. The new member is named for Sarasota County." The APAC pit is located in Section 12, T. 36 S., R. 18 E., Sarasota County. The quarry is now flooded as a lake in a Sarasota County park and can be accessed by scuba diving.

Age and Correlative Units. Based on the presence of several key chronostratigraphic index fossils, such as *Ecphora roxaneae* and *Chesapecten madisonius sarasotensis*, the Sarasota Member is now known to date from the late Zanclean Pliocene. The new member is exactly correlative with the upper beds of the Sunken Meadow Member of the Yorktown Formation, Chesapeake Group of Virginia and northern North Carolina, the basal beds of the Duplin Formation of southern North Carolina and northern South Carolina, and the Goose Creek Formation of southern South Carolina. The indurated layer at the top of the Sarasota Member most likely was produced during a time of prolonged subaerial exposure (and erosion) at the Zanclean–Piacenzian boundary.

Sarasota Member Index Fossils. Although the Sarasota sediments are leached and the majority of fossils are calcitic, many aragonitic species have survived dissolution. These are often in a poorly preserved, fragile state, with the exception being the large branches of the ahermatypic coral *Oculina sarasotana*. In the Quality Aggregates Pit #6 in Sarasota (separated from the area of the old APAC Pit by the I-75 highway), the Sarasota Member typically contained large reef-like clumps of the giant barnacle *Chesaconacavus tamiamiensis*, some of which reached 3 or 4 meters in length. These barnacle reefs interfingered with large bioherms composed of *Oculina sarasotana*, forming a distinctive ecosystem unlike any other known from the Neogene of Florida. Some of the principal Sarasota Member index fossils are shown on Figure 4.12.

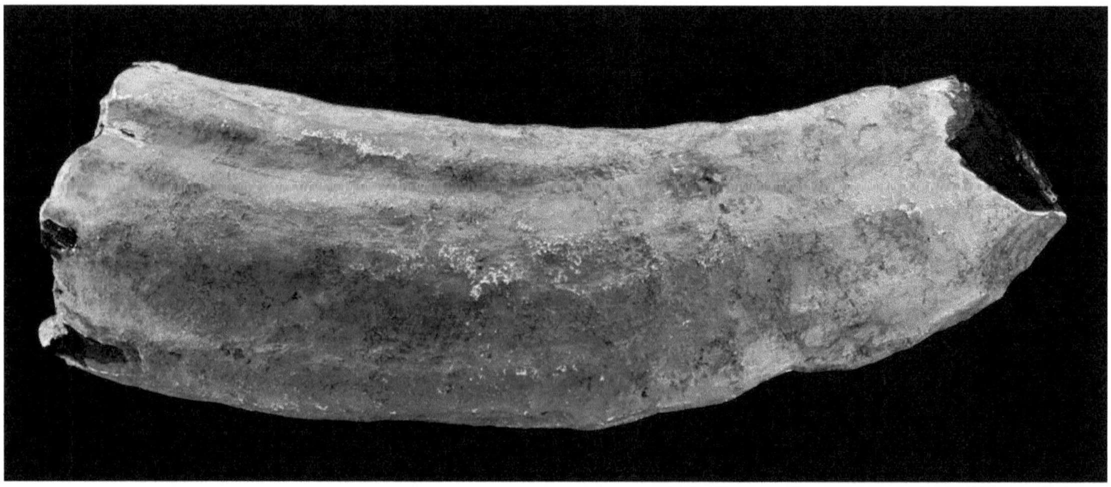

Florida Walrus (*Ontocetus emmonsi*) tusk (length 550 mm) from the Jupiter Waterway Member

Figure 4.12 Index Fossils of the Sarasota Member, Murdock Formation.
A= *Ecphora roxaneae* Petuch, 1991, length 102 mm; **B=** *Pterorhytis umbrifer* (Conrad, 1832), length 78 mm; **C=** *Tamianomia floridana* Mansfield, 1932, length 74 mm; **D=** *Chesapecten madisonius sarasotensis* Waller, 2018, 102 mm; **E=** *Planecphora hertweckorum* (Petuch, 1987), length 81 mm; **F=** *Chesaconcavus tamiamiensis* (Ross, 1965), length 152 mm; **G=** *Tropochasca* unnamed species; length 35 mm; **H=** *Scaphella kendrae* Petuch & Berschauer, 2021, holotype, length 43 mm (aperture filled with phosphatic sandstone). Figures G and H were collected in the indurated phosphatic limestone at the top of Unit 11 at Sarasota (referred to as "Unit 11A").

This immense pile of Pliocene fossils, composed mostly of mollusk shells, was excavated for road fill in the Quality Aggregates Pit #9 at Sarasota (with the senior author, in a red jacket, shown for scale). These beautifully-preserved specimens, all from the Tamiami Formation, readily demonstrate the abundance of fossils available to researchers in southern Florida and the Everglades region. Photo by Brian Schnirel.

A 98 mm specimen of the rare and seldom-seen *Latecphora violetae* (Petuch, 1988) from the Buckingham Member, Tamiami Formation in Quality Aggregates Pit #6 at Sarasota, Florida. This beautiful and unusual species has also been collected in the Goose Creek Formation of South Carolina.

CHAPTER 5. The Tamiami Subsea (Tamiami Formation: Buckingham, Pinecrest, and Ochopee Members)

The middle and late Pliocene (early and late Piacenzian Age) saw the return of extended periods of warm, tropical climates and biogenic carbonate deposition on the Florida Platform. By the early Piacenzian, permanent upwelling systems had ceased off western Florida, and the Okeechobean Sea no longer received massive influxes of plankton blooms and nutrients. This oceanographic shift permanently altered the Okeechobean depositional environments. The corresponding change in water chemistry resulted in the cessation of phosphatic sedimentation and the initiation of pure carbonate and carbonate-quartz sand deposition. With the loss of upwelling-influenced water conditions in the early Piacenzian, the Upwelling-Deltaic Depositional Episode came to an end and the Pseudoatoll Depositional Episode was ushered in. This marks the beginning of the deposition of the **Okeechobee Group** (discussed earlier in Chapter 1). At the end of Murdock deposition, a major regression caused the entire Okeechobean Sea to become emergent for over 200,000 years (Vail and Hardenbol, 1979; Haq *et al.*, 1988; Dowsett and Cronin, 1990). During this time, subaerial exposure caused the upper surface of the Sarasota Member to indurate into a thin layer of sandy phosphatic limestone.

Following the latest Zanclean–earliest Piacenzian sea level drop, the Okeechobean Sea basin was again reflooded, but this time by deeper and warmer water (30 m above present sea level: Brooks, 1974; Dowsett and Cronin, 1990). This marine transgression, which may have resulted from an extended period of global warming, produced the **Tamiami Subsea** (named for the Tamiami Formation) (Figure 5.1). During its span of over 2 million years, the Tamiami Subsea flooded four times, with the first two transgressions producing the deepest water conditions and lasting the longest (Figure 5.2). The last two transgressive intervals were much shorter in duration and were separated by longer periods of sea level lows. Judging from the gradual shallowing of each successive transgression, these last two Tamiami refloodings indicate the beginnings of global cooling and increased glaciation, presaging the Pleistocene. The eustatic fluctuations of the Tamiami Subsea are reflected in the four major depositional pulses seen in the Tamiami Formation, its principal stratigraphic unit. These led to the deposition of three sets of lithostratigraphic members, which are described in the following sections.

Deposition within the Tamiami Subsea, Okeechobean Sea. The oyster–barnacle banks and deltas of Hawthorn Group time formed the base for the classic late Pliocene structural feature of the Okeechobean Sea, the Everglades Pseudoatoll (see Petuch, 1987, 1997, 2004; Petuch and Roberst, 2207). This immense coral structure and its associated carbonate environments grew upon, and expanded, the geomorphologic features formed during the late Miocene and early Pliocene. At its maximum development during late Piacenzian time, the **Everglades Pseudoatoll** was a "U"-shaped feature composed of zonated coral reef systems surrounding a broad, shallow carbonate platform covered with Turtle Grass beds, open sand areas, and patch reef coral bioherms. The reef systems extended from Charlotte County, around the periphery of the Tamiami Subsea, all the way northward to Palm Beach County, making this the largest known Pliocene coral reef structure. If stretched out straight, the "U"-shaped pseudoatoll would have extended over 330 km, making it longer than the Great Barrier Reef of the Recent Belize coastline. The Everglades Pseudoatoll, then, is the largest contiguous coral reef system known from the Cenozoic western Atlantic. The geomorphology, depositional environments, and paleontology of the Everglades Pseudoatoll are given in Chapter 6.

Because of the structural complexity of the pseudoatoll and its proximity to the Floridian mainland, the Tamiami Subsea contained more types of tropical environments than were ever seen previously at any time in Florida's history. The main paleogeographical feature, the pseudoatoll, contained the highest number of depositional environments and ecological zones. Within the interior of the pseudoatoll, the dominant feature was the **Hendry Platform** (named for Hendry County; see Figure 6.1, Chapter 6). Unlike the surrounding coral reef tracts, which grew on late Miocene Hawthorn Formation oyster banks, the Hendry Platform developed from marine sediments that covered the surface of the drowned Immokalee Delta (Chapter 4). Sedimentation on the platform was complex, with the northern side receiving a large quartz sand component that derived from rivers draining off the peninsular highlands. The southern end of the Hendry Platform, being far from fluvial influences, received fewer siliciclastics and retained a higher percentage of carbonate sand and mud. The carbonate sedimentary input derived primarily from fines eroding off the Immokalee and Miami Reef Tracts (see Chapter 6). This lithofacies difference is readily seen in the contemporaneous Pinecrest and Ochopee Members of the Tamiami Formation (discussed later in this chapter). Here, the quartz sand-rich Pinecrest Member was deposited in the northern and central areas of the bank, whereas the carbonate-rich, low quartz-sand Ochopee was deposited in the extreme south.

During Tamiami time, the southern end of Florida was split into two smaller peninsulas, which were separated by a wide, shallow embayment. The western peninsula, the **Polk Peninsula** (named for Polk County) was the widest and was fringed by extensive mangrove forests and estuarine environments. The largest of these was the **Myakka Lagoon System** (named for the Myakka River, Sarasota County), which housed the richest Pliocene tropical estuarine fauna known from anywhere in the western Atlantic. A similar system of estuarine environments existed within the large bay-like feature east of the Polk Peninsula, the **Kissimmee Embayment** (named for the Kissimmee River). The Kissimmee Embayment differed from the Myakka Lagoon System in having much larger and more extensive areas of open, clean sand flats (both estuarine systems are discussed in detail in Chapter 7). To the east of the Kissimmee Embayment, the smaller **St. Lucie Peninsula** (named for St. Lucie County) separated the Okeechobean Sea from the open Atlantic. The exceptionally diverse environments of the Tamiami Subsea created habitats for an

Figure 5.1 Simulated satellite image (altitude: 100 miles) of the Floridian Peninsula during the late Piacenzian Pliocene, showing the probable appearance of the **Tamiami Subsea** of the Okeechobean Sea. This was the time of the maximum development of the **Everglades Pseudoatoll**, seen here as the prominent "u"-shaped system of coral reefs that surrounds three sides of the subsea basin. The old Immokalee Delta was now covered with carbonate sediments and had developed into the wide **Hendry Platform**. The southern end of Florida was composed of two smaller peninsulas; the **Polk Peninsula** on the west and the **St. Lucie Peninsula** on the east, with the narrow, elongated **Kissimmee Embayment** having developed between them. River systems draining off the western side of the Polk Peninsula formed the **Myakka Lagoon System**.

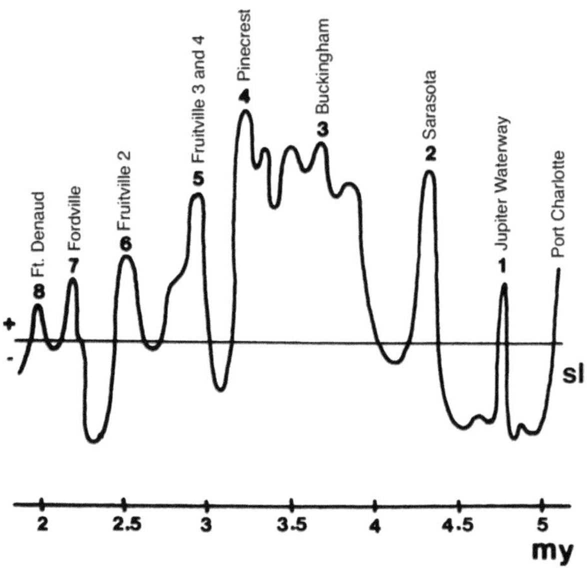

Figure 5.2 Sea level curve for the Pliocene, showing major eustatic fluctuations. Present sea level (sl) is used as a reference standard. Time is in millions of years (my). Numbers correspond to the times of deposition of formations and members, during sea level highs. For Formations: (1, 2) = **Murdock Formation**; (3, 4, 5, 6) = **Tamiami Formation**; (7, 8) = **Caloosahatchee Formation**. For Members: 1 = **Jupiter Waterway Member**, Murdock Formation; 2 = **Sarasota Member**, Murdock Formation; 3 = **Buckingham Member**, Tamiami Formation; 4 = **Pinecrest**, **Ochopee**, and **Golden Gate (basal beds) Members**, Tamiami Formation; 5 = **Fruitville** and **Golden Gate Members (Unit 3 and Unit 4 equivalents)**, Tamiami Formation; 6 = **Fruitville** and **Golden Gate Members (Unit 2 equivalent)**, Tamiami Formation; 7 = **Fordville Member**, Caloosahatchee Formation; 8 = **Fort Denaud Member**, Caloosahatchee Formation. (From the Oxygen Isotope Model of Krantz, D.E., 1991, Quaternary Science Reviews, 10: 163–174; Campbell, L.D., 1993, Pliocene Molluscs from the Yorktown and Chowan River Formations of Virginia, Virginia Division of Mineral Resources, Publication 127, Charlottesville, VA, 259 pp.)

unusually rich molluscan fauna that was replete with endemic local adaptive radiations. Within most units of the Tamiami Formation, the remnants of these unique faunas exist as beautifully preserved, thick shell beds, often filling entire exposed sections. Many of the more interesting and stratigraphically important of these mollusks are illustrated in the following sections of this chapter and also in Chapters 6 and 7.

The Tamiami Formation, Okeechobee Group. The Tamiami Formation, the principal stratigraphic unit of the Piacenzian Pliocene, has had a confused and ill-defined nomenclatural history. Because of the lack of surface exposures and because of the absence of collecting sites within the Everglades, the Tamiami Formation has had as many interpretations as there have been workers in southern Floridan geology. The name "Tamiami Limestone" was first proposed by Mansfield (1939, published posthumously) only for "… limestone penetrated in digging shallow ditches to form the road bed of the Tamiami Trail." This unit was later named the "Ochopee Limestone Member" by Muriel Hunter (1968) in her expanded nomenclatural scheme (discussed later in this section). In his 1939 paper, Mansfield also named the "Buckingham Limestone" from Lee County, considering it an older unit subjacent to his "Tamiami

Limestone." Based on geochronologic guide fossils, Mansfield's original assignment of an older, subjacent Buckingham and a younger, suprajacent "Tamiami" (Ochopee) is now known to be correct. Parker *et al.* (1955) were the first workers to attempt an expanded definition of the "Tamiami" as a formation.

In the early 1930s, Tucker and Wilson described the "**Acline Fauna**" from small shell pits near Punta Gorda (1932, 1933). Much of the Acline molluscan fauna was later encountered by Axel Olsson in excavations during construction of "Alligator Alley" and along the Tamiami Trail (U.S. Route 41) in the early 1960s. He later (Olsson *in* Olsson and Petit, 1964: 516) named the sandy units housing this Acline-type fauna the "Pinecrest Beds," after the old settlement of Pinecrest on the Everglades Road, which branches off U.S. Route 41 at the 40-Mile Bend on the Dade–Collier County line. He later recognized this same sandy unit and accompanying index fossils at other places around the Everglades, such as along the Miami Canal in western Broward and Palm Beach Counties, at the Brighton Indian Reservation on the northwestern side of Lake Okeechobee ("Brighton Facies"), and in canal digs in Glades, Highlands, and Hendry Counties. As more excavations occurred across the area in the 1970s and 1980s, geologists were given the opportunity to examine other facies of the Tamiami Formation. Some of these included coral reef tracts (the "Golden Gate Reef Member" of Missimer, 1992: 67; based on Meeder's unpublished dissertation) and lagoonal and estuarine facies (the "Fruitville Formation" of Waldrop and Wilson, 1990) from the Sarasota area. Earlier, Hunter (1968) expanded upon Parker's redefinition of the "Tamiami Formation," incorporating the Buckingham and Pinecrest with three new members, the previously mentioned Ochopee and two lower, phosphatic members, the Bayshore and Murdock Station. These last two "members" were removed from the Tamiami Formation and given full formational status by the senior author (Petuch, 2004; discussed in the previous chapter).

In 1982, the senior author published the first extensive measured section through a Tamiami ("Pinecrest Beds") exposure in the Ashland Petroleum and Asphalt Corp. (APAC) quarry in Sarasota. In that paper, 11 stratigraphic units were defined and numbered, with paleoenvironmental and biostratigraphic descriptions being given for each unit. Originally, three main units were recognized: an upper Caloosahatchee Formation (Unit 1), a middle "Pinecrest Formation" (Units 2–10), and a lower "Hawthorn Formation" (Unit 11). Since that time, with a greatly increased quality of geochronological, biostratigraphic, and paleoenvironmental data from around the Everglades area, we have been able to improve the original Sarasota stratigraphic column (shown here in Figure 5.3). As now understood, only Units 5–9 represent the Pinecrest, with Unit 11 representing the top of the Hawthorn Group (Sarasota Member of the Murdock Formation), Unit 10 representing the Buckingham, Units 2–4 representing the Fruitville, and Unit 1 representing the Fordville Member of the Caloosahatchee Formation (see Chapter 8). This section can be used as a standard reference section for the Pliocene of Florida and the Atlantic Coastal Plain. Of all the units and faunizones described by previous workers, we recognize only five nomenclatural subdivisions for the Tamiami Formation: the lower Buckingham Member, the middle Pinecrest and Ochopee Members, and the upper Fruitville and Golden Gate Members. Our definition is a compromise (updated with new and more accurate data) between the schemes of Hunter (1968), Waldrop and Wilson (1990), Lyons (1991), Missimer (1992), and Petuch and Roberts (2007).

Lithologic Description of the Tamiami Formation. Following Hunter's and Missimer's descriptions, with modifications, we here define the Tamiami Formation as consisting of "...variably interbedded calcarenites, calcilutites, quartz sands, and limestones and being devoid of massive phosphate deposition (except for reworked Murdock grains found in some facies of the Buckingham Member). Typically, many of the unconsolidated beds contain dense accumulations of bioclasts, mostly large mollusks and corals. Many of these shell beds, especially those composed mainly of bivalves, are so closely packed and dense that there is only a tiny fraction of any type of sediment matrix present between bioclasts. Clays are rarely present, being confined primarily to shoreline and estuarine facies, and always in minor quantities. The calcilutites are typically white or light gray, whereas the calcarenites and limestones are generally light gray, often with variable amounts of quartz sand intermixed. Quartz sand beds vary in color from white to dark gray, with some being stained yellow, brown, or light yellow-orange. The coral and mollusk bioclasts, when in fresh condition, are generally extremely well preserved and are usually stained a brown, tan, or bluish-gray color. Mollusk specimens from reducing environments in estuarine facies are stained black or dark blue-gray. Both aragonitic and calcitic fossils are typically present, although some limestone units are leached and contain only calcitic specimens." (Petuch and Roberts, 2007: 74-75).

The Tamiami Formation underlies the entire Everglades region (best seen in the distributional map of the Pinecrest and Ochopee Members), extending from Pinellas County in the northwest to Monroe County in the southwest and from southernmost Osceola County (Okeechobee Plain) south to Dade County. Throughout its entire range, there are no natural surface outcrops of the Tamiami. The formation ranges in thickness from 15 m at Sarasota, Sarasota County, to over 37 m at East Naples, Collier County (Missimer, 1992: 87–89), to over 60 m under the Atlantic Coastal Ridge (Swayze and Miller, 1984), to only 7 m in the Kissimmee Valley near Fort Drum, Okeechobee County. The surface contours of the Tamiami Formation follow the bathymetric contours of the Tamiami Subsea.

Stratotype. Although the name "Tamiami" has been used for this completely redefined and expanded formation, it is still necessary to give Mansfield's original type locality (from Mansfield, 1939: 8): " ... a limestone penetrated in digging shallow ditches to form the road bed of the Tamiami Trail over a distance of 34 mi in Collier and Monroe Counties, Florida." Because this section is really representative of only one of the five members of the Tamiami Formation (the Ochopee Member), we suggest that the measured section from the APAC pit in Sarasota (Figure 5.3) be added to Mansfield's stratotype as a standard reference section.

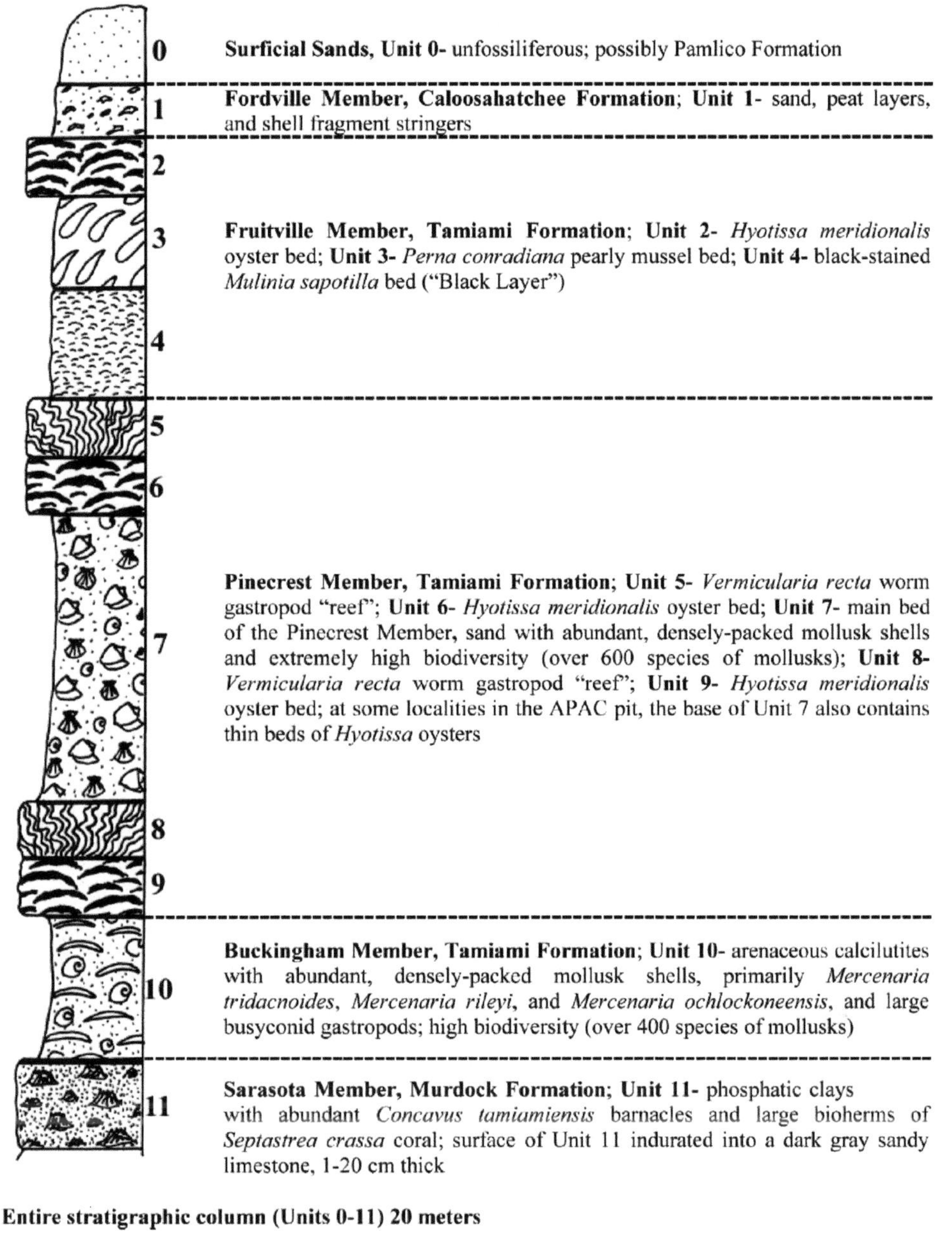

Figure 5.3 Stratigraphic column from the eastern side of the APAC pit (formerly Richardson Quarry and Warren Brothers pit), Fruitville area of Sarasota, Sarasota County, Florida. Three formations were exposed within the quarry; the **Caloosahatchee Formation** (Fordville Member, Unit 1), the **Tamiami Formation** (Fruitville, Pinecrest, and Buckingham Members, Unit 2 to Unit 10), and the **Murdock Formation** (Sarasota Member, Unit 11). A Unit 12 (Murdock Formation, Jupiter Waterway Member) was later exposed at the bottoms of deep drainage channels (gray phosphatic clays with dense beds of large scallops). Unit 11 at this quarry is the stratotype of the **Sarasota Member** of the Murdock Formation. (Modified from Petuch, E.J., 1982, Proceedings of the Academy of Natural Sciences of Philadelphia, 134: 12–30).

Age and Correlative Units. Based upon entire suites of geochronologic index fossils, the Tamiami Formation is now known to span most of the late Pliocene, from the earliest Piacenzian to the latest Piacenzian. The presence of diagnostic species of widespread molluscan groups, such as the bivalves *Carolinapecten*, *Nodipecten*, and

Mercenaria and the gastropods *Pliculofusus*, *Latecphora*, and *Contraconus* (discussed in the following sections), show that the formation is exactly correlative with the Rushmere, Mogarts Beach, and Moore House Member of the Yorktown Formation of Virginia and northern North Carolina, and the Edenhouse Member of the Chowan River Formation of the same area. The Tamiami is also correlative with the middle and upper beds of the Duplin Formation of southern North Carolina and northern South Carolina, the middle and upper beds of the Goose Creek Formation of South Carolina, and the Jackson Bluff Formation of the Florida Panhandle.

Tamiami Index Fossils. The single most important index fossil group within the Tamiami Formation is the gastropod family Cypraeidae, the cowrie shells. During the Piacenzian Pliocene, cowries underwent amazing species radiations within the Okeechobean Sea, producing whole suites of endemic, chronologically restricted species. There are 79 described species now known from the Tamiami Formation, encompassing 6 genera and 11 subgenera, making this the single largest species radiation of fossil cowries ever found in one place anywhere on Earth (Petuch, 1994, 1996, 1998, 2004; Petuch, Berschauer, and Myers, 2018). With the exception of two widespread eastern American species (*Akleistostoma carolinensis* and *A. pilsbryi*), all the members of these species radiations were confined to the area of the Tamiami Subsea of the Okeechobean Sea. The Florida cowries were gregarious, occurring in huge aggregations numbering in the tens of thousands (much like the living mouse cowrie *Muracypraea mus* of Venezuela, a distant relative with similar ecological preferences; see Petuch, 1976; 1979; 1988). Their abundance, large size, and rapid evolution make them ideal stratigraphic index fossils. All 79 described Tamiami species are illustrated in the following sections and are arranged in stratigraphic context. Other important Tamiami index fossil groups, which can be used to demarcate the boundaries between members, include the gastropod genera *Hystrivasum*, *Pliculofusus*, *Echinofulgur*, *Tropochasca*, and *Pterorhytis* and the bivalve genera *Carolinapecten* and *Nodipecten*.

Buckingham Member, Tamiami Formation. Although originally recognized by Matson and Clapp (1909) and Cooke and Mossom (1929), the outcrops of limestone exposed in the Caloosahatchee River channelizing excavations and in quarries near Buckingham, Lee County, were assigned to the "Choctawhatchee Marl." It was not until 1939 that Mansfield formally named the unit (as the "Buckingham Limestone") and gave a detailed list of molluscan index fossils. While the paper was in press (published September 1, 1939), Wendell Mansfield died (July 24) having seen the galley proofs before his death but not the final page proofs. Later, Cooke (1945: 210) changed the name to "Buckingham Marl," basing this refinement on work that he had done a year earlier with Gerald Parker. They noticed that the exposed, weathered surface indurates into a limestone but that the unexposed material is actually an unconsolidated fossiliferous calcarenite. Hunter (1968) placed the Buckingham within her expanded concept of the Tamiami Formation, in this case as the "Buckingham Limestone

Member." Based on the misidentification of some scallop species (*Nodipecten* and the *Christinapecten tamiamiensis* species complex), Hunter considered the Buckingham to be time-equivalent with the Ochopee and Pinecrest Members. Missimer (1992: Figure 3.3) correctly identified the Buckingham as being subjacent to the Ochopee Member and as being the oldest and lowest part of the Tamiami sequence.

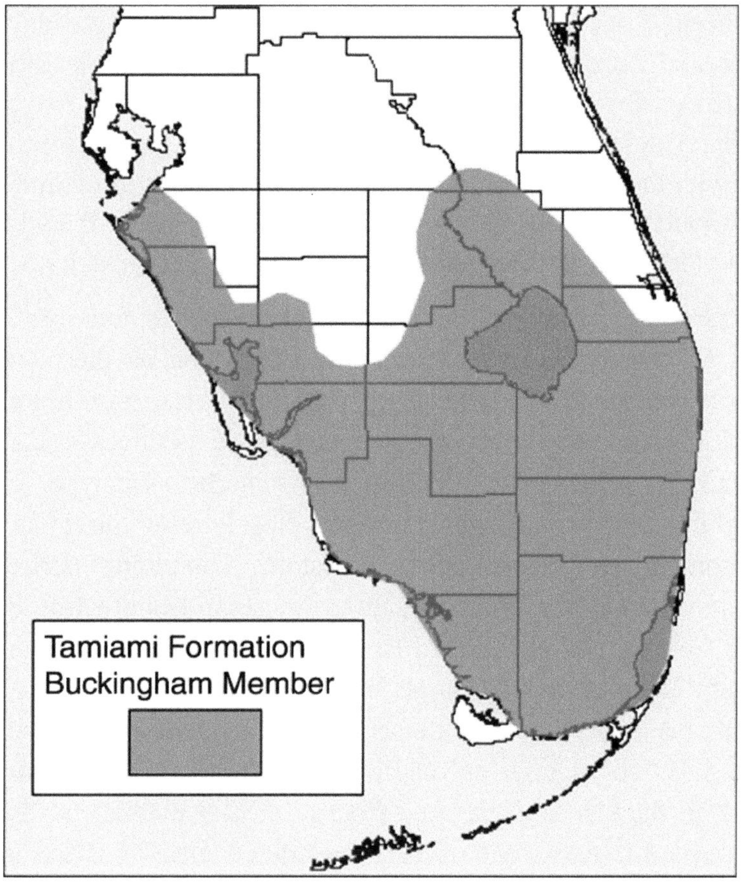

Figure 5.4 Areal distribution of the Buckingham Member, Tamiami Formation in southern Florida.

Lithologic Description and Areal Distribution. The Buckingham Member was originally only sketchily described (from Mansfield, 1939: 8, 12): "A new formation is here proposed for a limestone cropping out in Lee County, Florida … The molluscan fauna of the Buckingham limestone consists mainly of *Pecten* and *Ostrea*, which are well-preserved; but most of the other genera are preserved only as casts and molds … The matrix in which the fossils are embedded consists of a chalky limestone that contains little sand and many small grains of brown phosphorite. The rock hardens on exposure and changes to a brownish color." Cooke (1945: 210) further expanded the lithostratigraphic definition: "Fresh exposures of the Buckingham marl consist of cream-colored impermeable calcareous clay containing small grains of a brown phosphatic mineral, which increase in abundance toward the lower part of the formation. The Buckingham is fairly uniform throughout its known extent, but where it has long been exposed to weathering, the marl hardens into firm limestone." Hunter's description essentially follows Cooke's; "Light gray to white soft calcareous clay (a calcilutite) that weathers to a buff color. It contains some quartz sand, a few bone fragments and shark teeth, and few grains of brown phosphate. Poorly preserved fossil molds and well-preserved pectens, oysters, barnacles, echinoids, etc., are present." The most interesting aspect that is mentioned in all three descriptions is the presence of a "brown phosphorite." As Missimer (1992: 68) pointed out, these phosphatic mineral grains are reworked from the underlying Hawthorn Group. These would have eroded off the indurated surface of the Murdock Formation when the Everglades area was reflooded by the Tamiami Subsea. These brown (and etched) phosphate grains were later mixed in with the younger Buckingham sediments. The gradational content of the phosphate component pointed out by Cooke further indicates that these grains were reworked from older beds and were not part of Buckingham deposition.

The best and largest exposures of the Buckingham Member were in the Quality Aggregates pit #6 and at the APAC pit, both at Sarasota. At these two localities, the Buckingham was composed of a sandy lime mud (arenaceous calcilutite) with abundant, closely packed mollusk shell bioclasts, mostly large bivalves such as *Mercenaria tridacnoides* and *Carolinapecten walkerensis*. As at the stratotype, exposed surfaces often weathered into a soft arenaceous limestone, closely resembling that at Buckingham. Blocks of a non-phosphatic arenaceous limestone, containing perfect specimens of *Mercenaria tridacnoides*, were also brought up by draglines at Mule Pen Quarry, East Naples, Collier County, and at the Bird Road lake excavation in Miami, Dade County. The Buckingham Member has the same areal distribution as does the entire Tamiami Formation, with the exception of Pinellas County, where the member appears to be absent (Figure 5.4). Throughout its distribution, the member is variable in thickness, ranging from 7 m thick near Alva, Lee County, to 2.5 m thick at Sarasota, Sarasota County, and to over 8 m thick west of Miami, Dade County. The surface contours of the Buckingham Member follow the bathymetric contours of the Tamiami Subsea.

Stratotype. From Mansfield (1939: 8): " ... The type locality is at a quarry near State Highway 25, half a mile west of Orange River, Lee County, Florida (Sec. 5, T. 44 S., R. 26 E.)." The APAC section (Figure 5.3) can be used as a standard reference section for the Buckingham, showing the suprajacent and subjacent units.

Age and Correlative Units. Based on the presence of key stratigraphic index fossils, such as *Akleistostoma carolinensis*, *Planecphora mansfieldi*, *Mercenaria tridacnoides*, *Nodipecten peedeensis*, and *Lirophora ulocyma*, the Buckingham Member is now known to date from the earliest Piacenzian Pliocene. The member is exactly correlative with the Rushmere Member of the Yorktown Formation of Virginia and northern North Carolina, the lower beds of the Duplin Formation of southern North Carolina and northern South Carolina, the lower beds of the Goose Creek Formation of South Carolina, and the lower member (unnamed, the "*Ecphora* Zone" at Alum Bluff) of the Jackson Bluff Formation of the Florida Panhandle (Mansfield, 1930).

Figure 5.5 Cowries of the Buckingham Member, Tamiami Formation.
A, B= *Akleistostoma (Akleistostoma) carolinensis* (Conrad, 1841), length 60 mm; **C, D=** *Akleistostoma (Mansfieldicypraea) crocodila* (Petuch, 1994), length 64 mm; **E, F=** *Akleistostoma (Ingramicypraea) pilsbryi* (Ingram, 1947), length 26 mm; **G, H=** *Siphocypraea (Seminolecypraea) micanopy* Petuch and Drolshagen, 2011, length 58 mm; **I, J=** *Pahayokea (Gardnericypraea) erici* (Petuch, 1998), length 74 mm.

Figure 5.6 Gastropods of the Buckingham Member, Tamiami Formation.
A, B= *Calusacypraea duerri* (Petuch, 1996), length 65 mm; **C, D**= *Pseudadusta buckinghamensis* Petuch and Drolshagen, 2011, length 53 mm; **E**= *Macrostrombus briani* Petuch and Drolshagen, 2011, length 173 mm; **F**= *Strombus floridanus* Mansfield, 1930, length 68 mm; **G**= *Ecphora quadricostata striatula* Petuch, 1986, length 57 mm (also found in the Jackson Bluff Formation at Alum Bluff, northern Florida); **H**= *Planecphora mansfieldi* (Petuch, 1989), length 85 mm (also found in the Jackson Bluff Formation at Alum Bluff, northern Florida); **I**= *Hystrivasum violetae* Petuch, 1994, length 67 mm; **J**= *Apertifusus dianeae* (Petuch, 1994), length 135 mm.

Figure 5.7 Gastropods of the Buckingham Member, Tamiami Formation.
A= *Tropochasca petiti* Olsson, 1967, length 85 mm; **B=** *Echinofulgur cannoni* Petuch, 1994, length 69 mm; **C=** *Phyllonotus leonensis* (E. Vokes, 1967), length 75 mm; **D=** *Pterorhytis marshalli* (Mansfield, 1930), length 42 mm; **E=** *Chicoreus riparius* E. Vokes, 1967, length 37 mm; **F=** *Busycon alumensis* Mansfield, 1930, length 75 mm (also found in the lower beds of the Jackson Bluff Formation); **G=** *Busycon filosum* (Conrad, 1862), length 130 mm; **H=** *Busycon pachyus* Petuch, 1994, length 152 mm; **I=** *Sinistrofulgur hollisteri* Petuch, 1994, length 187 mm; **J=** *Brachysycon propeincile* (Mansfield, 1930), length 68 mm (also found in the lower beds of the Jackson Bluff Formation of northern Florida).

Figure 5.8 Gastropods of the Buckingham Member, Tamiami Formation.
A= *Sinistrofulgur grabaui* Petuch, 1994, length 193 mm; **B**= *Lindafulgur lindajoyceae* (Petuch, 1991), length 114 mm; **C**= *Busycoarctum tudiculatum* (Dall, 1890), length 94 mm; **D**= *Pyruella harasewychi* Petuch, 1982, length 73 mm; **E**= *Pyruella waltfrancei* Petuch, 1994, length 60 mm; **F**= *Pyruella streami* Petuch, 1994, length 56 mm; **G**= *Fulguropsis carolinensis* (Tuomey and Holmes, 1856), length 121 mm; **H**= *Busycotypus libertiense* (Mansfield, 1930), length 164 mm; **I**= *Scaphella martinshugari* Petuch, 1994, length 59 mm; **J**= *Volutifusus emmonsi* Petuch, 1994, length 110 mm; **K**= *Hystrivasum barkleyae* Petuch, 1994, length 85 mm.

Figure 5.9 Mollusks of the Buckingham Member, Tamiami Formation.
A= *Scaphella ashleyae* Petuch, 1994, length 115 mm; **B**= *Contraconus lindajoyceae* Petuch, 1991, length 87 mm; **C**= *Americoliva keatoni* Petuch, 1994, length 61 mm; **D**= *Kohniconus patstreamae* (Petuch, 1994), length 78 mm; **E**= *Seminoleconus violetae* (Petuch, 1991), 53 mm; **F**= *Osceolaconus buckinghamensis* Petuch and Drolshagen, 2011, length 61 mm; **G**= *Lindaconus ductor* Petuch and Drolshagen, 2011, length 54 mm; **H**= *Mercenaria tridacnoides* (Lamarck, 1818), length 122 mm; **I**= *Nodipecten peedeensis* (Tuomey and Holmes, 1856), length 138 mm; **J**= *Ptychosalpinx perprotractus* Petuch, 1991, length 62 mm; **K**= *Volutifusus dougsheltoni* Petuch and Berschauer, 2021, length 120 mm (lowest Buckingham beds only).

Buckingham Index Fossils. The Buckingham molluscan fauna is one of the richest known from the Okeechobee Group. The highly distinctive and characteristic assemblages found in this member contain three faunal components: an endemic Okeechobean Sea component, a component of species shared with the "*Ecphora* Zone" of the Jackson Bluff Formation (the Jackson Subsea of the Choctaw Paleosea; see Petuch, 2004), and a component of widespread species shared with the Duplin and Goose Creek Formations. The marine communities of the Buckingham Member were described by the senior author (Petuch, 2004: 166–171) and include the ***Mercenaria tridacnoides*** **Community** (shallow muddy sand areas and intertidal mud flats), the ***Apicula buckinghamensis*** **Community** (shallow lagoon turritella beds), the ***Nodipecten peedeensis*** **Community** (deep lagoon pectinid beds), and the ***Phyllangia blakei*** **Community** (lagoonal coral bioherms). Several ahermatypic corals are also shared with the Jackson Bluff Formation (illustrated in Petuch, 2004: Plate 54) and include *Astrangia leonensis*, *Phyllangia blakei*, and *Paracyathus vaughani*. Some of the more abundant and indicative index fossils for the Buckingham Member are shown in the preceding iconography.

Pinecrest Member, Tamiami Formation. When Olsson (in Olsson and Petit, 1964: 516–518) formally proposed the name "Pinecrest Beds," he actually defined a molluscan biozone and not a lithostratigraphic unit. This was later redescribed and restricted by Hunter (1968: 443) in her expanded definition of the Tamiami Formation. In that work, Hunter named the unit the "Pinecrest Sand Member," and gave the first lithologic description of Olsson's widespread shell beds and their characteristic fauna. A more detailed description of the Pinecrest, showing a wider range of lithologies and paleoenvironments, was given by the senior author (Petuch, 1982) in a study of the APAC quarry at Sarasota (see Figure 5.3). Later, Missimer (1992: 65) gave more details regarding the areal distribution of what he called the "Pinecrest Sand Member." In this book, as in the previous Everglades geology book (Petuch and Roberts, 2007), we refer to this middle Tamiami unit as simply the Pinecrest Member. To paleontologists, the molluscan fauna of the Pinecrest Member is particularly noteworthy in that it is the richest-known in the Okeechobee Group and the entire Pliocene of the Atlantic Coastal Plain. These mollusks lived during the warmest and most tropical time of the entire Pliocene (Figure 5.2), and over 1000 species have been documented from this single unit alone (Petuch, 1994). This number may rise to well over 1500 when the micromolluscan fauna is better studied.

Lithologic Description and Areal Distribution. As pointed out by Hunter (1968), the Pinecrest Member is composed of well-sorted, clean quartz sand, locally quite shelly, with abundant well-preserved marine fossils. From the stratigraphic column and descriptions shown in Figure 5.3, some of the Pinecrest shell beds are so dense and tightly packed that only a small amount of quartz sand exists between the large molluscan bioclasts. Some of these shell beds are essentially monocultures, being made up almost entirely of either *Hyotissa meridionalis* oysters, *Carolinapecten eboreus* scallop beds, *Vermicularia recta* worm gastropods, or the small strombid gastropod

Strombus fruitvillensis. The quartz sand component is usually white, pale yellow, or light gray in color. Variable amounts of lime muds (calcilutites) are often mixed with the quartz sand. The calcilutite/quartz sand ratio is gradational, with a higher quartz sand content in areas adjacent to the paleopeninsular mainland and a lower quartz sand content in areas adjacent to the pseudoatoll reefs and the Loxahatchee Trough. The shell bioclasts, when fresh, are stained a cream, tan, or light gray color.

The areal extent of the Pinecrest Member is equivalent to that of the entire Tamiami Formation, except for southern Dade and northern Monroe Counties, where it interfingers with the contemporaneous Ochopee Member, and under the Immokalee Rise and the Atlantic Coastal Ridge, where it interfingers with the reefal deposits of the lower part of the Golden Gate Member. Thicknesses of the Pinecrest Member are variable, ranging from 10 m at Sarasota, to 2 m in southern Lee County, to over 7 m at Fort Drum, Okeechobee County, to over 20 m west of the Atlantic Coastal Ridge in Palm Beach County. The surface contours of the Pinecrest Member conform to the bathymetric contours of the Tamiami Subsea. Within the Okeechobee Plain and Kissimmee River Valley, the lithology of equivalent beds differs from other areas of Pinecrest deposition. Here, high percentages of mud, clay minerals, and particulate organic matter are mixed with the quartz sand and shell bioclast components. This separate, distinct lithology, here referred to as the **Kissimmee Facies** of the Pinecrest Member, formed within the extensive mangrove forest and mud flat environments of the mid-Piacenzian Kissimmee Embayment. The unique environments of this area housed a highly endemic molluscan fauna, many of which are still unnamed and unstudied.

Stratotype. Olsson (1964: 516) notes that the Pinecrest Member is named "… for certain strata composed largely of sand, barren or highly fossiliferous, encountered directly below a surface limestone in the general region of the 40-Mile Bend on the Tamiami Trail (Route 41) west of Miami in the western part of Dade County and extending across its boundary into Collier County, Florida. The name is taken from an old settlement on the Everglades Road (which branches off the present highway at 40-Mile Bend) about 1 mi west of the Dade–Collier County line." The APAC section from Sarasota (Figure 4.3) can be used as a standard reference section for the Pinecrest Member, showing the suprajacent and subjacent beds.

Age and Correlative Units. Based on the presence of key widespread stratigraphic molluscan index fossils, such as *Latecphora bradleyae*, *Carolinapecten eboreus*, *Argopecten comparilis*, *Contraconus adversarius*, and *Pliculofusus sparrowi*, the Pinecrest Member is now known to date from the mid-Piacenzian Pliocene. The member is exactly correlative with the Mogarts Beach Member of the Yorktown Formation, Chesapeake Group of Virginia and northern North Carolina, the main beds of the Duplin Formation of southern North Carolina and northern South Carolina, the upper beds of the Goose Creek Formation of South Carolina, and the upper (unnamed) member (the "*Cancellaria* Zone") of the Jackson Bluff Formation of the Florida Panhandle (Mansfield, 1930, 1932).

Figure 5.10 View of an exposure of the Pinecrest Member, Tamiami Formation, Unit 7 in the APAC Pit, Sarasota, Sarasota County, Florida. A large specimen of the fasciolariid gastropod, *Triplofusus duplinensis*, protrudes from the wall to the right of the hammer. The scallop *Carolinapecten eboreus* and the gryphaeid oyster *Hyotissa meridionalis* are abundant in this exposure, as is the small strombid gastropod *Strombus fruitvillensis*.

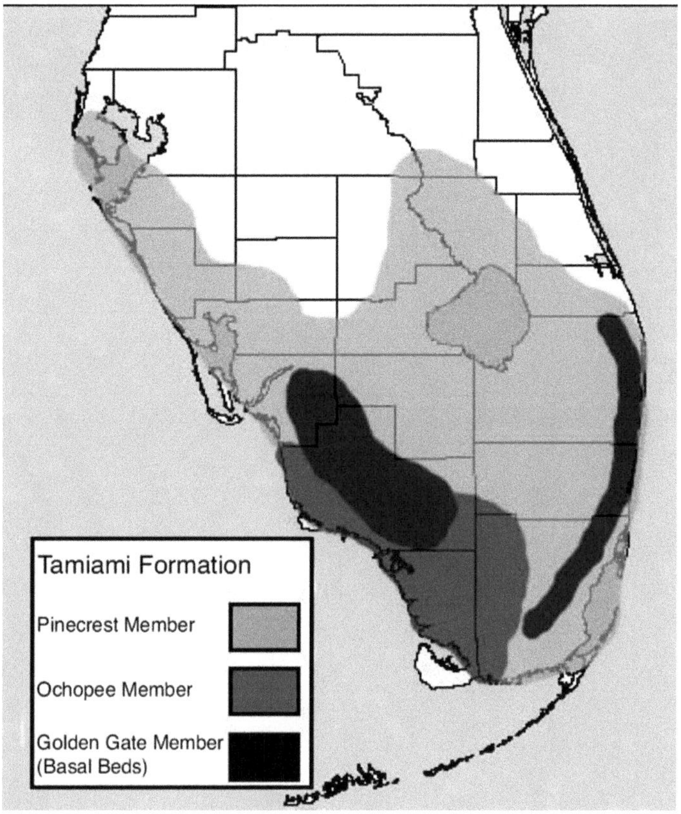

Figure 5.11 Areal distribution of the Pinecrest, Ochopee, and Golden Gate (basal beds) Members of the Tamiami Formation in southern Florida.

Pinecrest Index Fossils. Unlike the fauna of the Buckingham Member, which contains many lower Jackson Bluff taxa, the Pinecrest fauna contains a much higher percentage of endemic taxa, approaching 60%. The other components comprise species shared with the Yorktown, Duplin, and upper beds of the Jackson Bluff Formation. The marine communities of the Pinecrest Member were described by the senior author (Petuch, 2004: 171–177; for both the Myakka Lagoon and the Kissimmee Embayment). During Pinecrest time, the Kissimmee Embayment area contained a molluscan fauna that was very different from that of the Myakka Lagoon area, with a high degree of endemism being present in both areas and with very few taxa in common. These faunal differences were the result of very different habitats in the two main estuarine areas (discussed in Chapter 7; see Petuch, 2004; Petuch and Roberts, 2007; Petuch, Berschauer, and Myers, 2018). Some of the most unusual and exotic fossil shells found anywhere in Florida have been described from the Pinecrest Member, and examples of these are shown here in the following iconography.

Ochopee Member, Tamiami Formation. When the Tamiami Trail ("Tampa to Miami", now U.S. Route 41) was being constructed in the late 1920s and early 1930s, a hard, distinctive limestone was exposed in shallow excavations on the southwestern side of the Everglades. This was later named the Tamiami Limestone by Mansfield (1939). The Tamiami material was extensively excavated, crushed, and used as the

Vermicularia recta (Olsson and Harbison, 1953), aggregation length 155 mm.

Mansfield was able to study the fossil fauna in depth, describing several new species. After more research was undertaken in the 1950s and 1960s (by Parker, Puri, Vernon, and Olsson; all cited previously in Chapter 1), it was found that Mansfield's bed for the new trans-Florida highway. Because of the abundance of exposures, "Tamiami Limestone" was simply a facies of a much larger, lithologically complex unit. Preserving the well-known name "Tamiami," Hunter (1968: 443) expanded the formational concept to include the other contemporaneous lithologic units around the Everglades area, naming the entire suite the "Tamiami Formation." She then renamed Mansfield's original "Tamiami Limestone" as the "Ochopee Limestone Member." Although this was not in complete concordance with the Code of Stratigraphic Nomenclature, Hunter's scheme has been accepted by consensus, and we follow her nomenclature in this book, referring to this unit as the Ochopee Member of the Tamiami Formation.

Figure 5.12 Cowries of the Pinecrest Member, Tamiami Formation.
A, B= *Akleistostoma floridana* (Mansfield, 1931), length 66 mm; **C, D**= *Akleistostoma (Paludacypraea) patula* Petuch and Drolshagen, 2011, length 62 mm; **E, F**= *Akleistostoma (Mansfieldicypraea) rugostoma* Petuch and Drolshagen, 2011, length 78 mm; **G, H**= *Siphocypraea (Seminolecypraea) trippeana* Parodiz, 1988, length 57 mm; **I, J**= *Pahayokea (Gardnericypraea) mansfieldi* (Petuch, 1998), length 76 mm.

Figure 5.13 Gastropods of the Pinecrest Member, Tamiami Formation.
A, B= *Calusacypraea globulina* Petuch, 2004, length 55 mm; **C, D**= *Pseudadusta hertweckorum* (Petuch, 1991), length 57 mm; **E**= *Macrostrombus hertweckorum* (Petuch, 1991), length 176 mm; **F**= *Cassis floridensis* Tucker and Wilson, 1932, length 124 mm; **G**= *Linatella floridana* (Mansfield, 1930), length 72 mm; **H**= *Latecphora bradleyae* (Petuch, 1988), length 68 mm; **I, J**= *Calusacypraea (Myakkacypraea) briani* (Petuch, 1996), length 72 mm; **K**= *Latecphora bradleyae* (Petuch, 1988), length 93 mm, with *Balanus newburnensis* barnacles growing on its spire whorls.

Figure 5.14 Muricid Gastropods of the Pinecrest Member, Tamiami Formation.
A= *Pterorhytis fluviana* (Dall, 1903), length 53 mm; **B=** *Ecphora quadricostata rachelae* Petuch, 1988, length 63 mm (uppermost beds of the Pinecrest Member only); **C=** *Hexaplex hertweckorum* (Petuch, 1988), length 80 mm; **D=** *Vitularia linguabison* E. Vokes, 1967, length 49 mm; **E=** *Subpterynotus feliciae* Petuch and Berschauer, 2020, length 49 mm; **F=** *Murexiella faceta* (E. Vokes, 1963), length 25 mm; **G=** *Vokesinotus lamellosus* (Emmons, 1858), length 32 mm; **H=** *Eupleura metae* Petuch, 1994, length 37 mm; **I=** *Eupleura brevispira* Mansfield, 1930, length 27 mm; **J=** *Latecphora bradleyae hertwecki* (Petuch, 1989), length 30 mm (uppermost beds of the Pinecrest Member only); **K=** *Aspella petuchi* E. Vokes, 1995, length 13 mm.

Figure 5.15 Gastropods of the Pinecrest Member, Tamiami Formation.
A= *Chicoreus stephensae* Petuch, 1994, length 75 mm; **B**= *Vokesimurex violetae* Petuch, 1994, length 42 mm; **C**= *Acantholabia sarasotaensis* Petuch, 1991, length 33 mm; **D**= *Chicoreus xestos* E. Vokes, 1974, length 38 mm; **E**= *Stramonita caloosana* (Tucker and Wilson, 1933), length 46 mm; F= *Phyllonotus globosus* (Emmons, 1858), length 79 mm; **G**= *Pyruella rugosicostata* Petuch, 1982, length 65 mm; **H**= *Pyruella seminole* Petuch, 1994, length 70 mm; **I**= *Chicoreus floridanus* E. Vokes, 1965, length 58 mm; **J**= *Cinctura rhomboidea* (Rogers, 1839), length 122 mm (also found in the Yorktown and Duplin Formations).

Figure 5.16 Gastropods of the Pinecrest Member, Tamiami Formation.
A= *Sinistrofulgur contrarium* (Conrad, 1840), length 72 mm; **B**= *Fulguropsis excavatum* (Conrad, 1840), length 102 mm; **C**= *Busycoarctum tropicalis* (Petuch, 1994), length 174 mm; **D**= *Busycotypus bicoronatum* (Tripp, 1988), length 88 mm; **E**= *Melongena taurus* Petuch, 1994, length 215 mm; **F**= *Tropochasca lindae* Petuch, 1994, length 107 mm; **G**= *Echinofulgur dalli* Petuch, 1994, length 100 mm; **H**= *Pyruella osceola* Petuch, 1982, length 112 mm; **I**= *Laevisycon demistriatum* (Petuch, 1982), length 60 mm; **J**= *Pyruella miccosukee* Petuch, 1991, length 55 mm.

Figure 5.17 Gastropods of the Pinecrest Member, Tamiami Formation.
A= *Pliculofusus sparrowi* (Emmons, 1858), length 148 mm; **B=** *Pliculofusus nodulosus* (Emmons, 1858), length 180 mm; **C=** *Calophos wilsoni* Allmon, 1990, length 68 mm; **D=** *Eichwaldiella magnasulcus* Petuch, 1991, length 113 mm; **E=** *Hystrivasum locklini* (Olsson and Harbison, 1953), length 96 mm; **F=** *Scaphella trenholmii* (Tuomey and Holmes, 1856), length 85 mm; **G=** *Scaphella mansfieldi* Petuch, 1994, length 71 mm; **H=** *Volutifusus spengleri* Petuch, 1991, length 198 mm; **I=** *Polygona noseli* Lyons, 1991, length 49 mm; **J=** *Ecphora quadricostata umbilicata* (Wagner, 1839), length 96 mm (lowest beds of the Pincrest Member only).

Figure 5.18 Gastropods of the Pinecrest Member, Tamiami Formation.
A= *Turbinella streami* Petuch, 1991, length 235 mm; **B**= *Americoliva carolinensis* (Conrad, 1863), length 85 mm; **C**= *Bullata popenoi* (Mansfield, 1930), length 63 mm; **D**= *Pleioptygma debrae* Petuch, 1994, length 118 mm; **E**= *Pleioptygma lindae* Petuch, 1991, length 119 mm; **F**= *Trigonostoma druidi* Olsson and Petit, 1964, length 36 mm; **G**= *Triplofusus duplinensis* (B. Smith, 1940), length 356 mm; **H**= *Cymatophos lindae* Petuch, 1991, length 69 mm; **I**= *Strombus fruitvillensis* Petuch and Drolshagen, 2011, length 66 mm.

Figure 5.19 Mollusks of the Pinecrest Member, Tamiami Formation.
A= *Contraconus adversarius* (Conrad, 1840), length 98 mm; **B=** *Herndliconus paranobilis* (Petuch, 1991) length 36 mm; **C=** *Dauciconus bassi* (Petuch, 1991), length 35 mm; **D=** *Jaspidiconus marymansfieldae* (Petuch, 1994), length 24 mm; **E=** *Lindaconus cherokus* (Olsson and Petit, 1964), length 73 mm; **F=** *Pyruconus druidi* (Olsson, 1967), length 147 mm; **G=** *Kohniconus stowelli* Petuch and Drolshagen, 2011, length 112 mm; **H=** *Seminoleconus trippae* (Petuch, 1991), length 68 mm; **I=** *Carolinapecten eboreus* (Conrad, 1833), length 86 mm; **J=** *Nodipecten collierensis* (Mansfield, 1931), length 92 mm, also found in the Ochopee Member.

Figure 5.20 View of a bioherm of the worm gastropod *Vermicularia recta*, intercalated between two layers of the gryphaeid oyster *Hyotissa meridionalis*, within the Pinecrest Member of the Tamiami Formation in the APAC Pit, Sarasota, Sarasota County, Florida. *Hyotissa meridionalis* is also abundant in the Ochopee Member.

Figure 5.21 Cowries of the Pinecrest Member (Kissimmee Facies), Tamiami Formation.
A, B= *Akleistostoma (Ingramicypraea) highlandsensis* Petuch and Drolshagen, 2011, length 31 mm; **C, D**= *Akleistostoma (Olssonicypraea) hughesi* (Olsson and Petiti, 1964), length 62 mm; **E, F**= *Pahayokea kissimmeensis* (Petuch, 1994), length 53 mm; **G, H**=*Pahayokea (Kissimmecypraea) transitoria* (Olsson and Petit, 1964), length 69 mm; **I, J**= *Okeechobea osceolai* Petuch and Drolshagen, 2011, length 51 mm.

Figure 5.22 Gastropods of the Pinecrest Member (Kissimmee Facies), Tamiami Formation.
A, B= *Okeechobea (Yeehawcypraea) bartoni* Petuch and Drolshagen, 2011, length 56 mm; **C**= *Brachysycon kissimmeensis* (Petuch, 1994), length 111 mm; **D**= *Pyruella carraheri* Petuch, 1994, length 63 mm; **E**= *Pterorhytis lindae* Petuch, 1994, length 47 mm; **F**= *Pterorhytis lindae* view of spire of same specimen; **G**= *Septa henicum* (Woodring, 1959), length 104 mm; **H**= *Chicoreus shirleyae* E. Vokes, 1974, length 80 mm; **I**= *Volutifusus obtusus* (Emmons, 1858), length 130 mm.

Figure 5.23 Gastropods of the Pinecrest Member (Kissimmee Facies), Tamiami Formation.
A= *Hystrivasum olssoni* (E. Vokes, 1966), length 107 mm; **B**= *Hystrivasum vokesae* (Hollister, 1971), length 75 mm; **C**= *Pliculofusus lindae* (Petuch, 1994), length 142 mm; **D**= *Pliculofusus osceolai* (Petuch, 1994), length 98 mm; **E**= *Osceolaconus matchetti* Petuch and Drolshagen, 2011, length 71 mm; **F**= *Kohniconus duerri* (Petuch, 1994), length 62 mm; **G**= *Ptychosalpinx kissimmeensis* Petuch, 1994, length 56 mm; **H**= *Monoplex valentinei* (Olsson and Petit, 1964), length 96 mm; **I**= *Busycotypus conradi* (Tuomey and Holmes, 1856), length 210 mm.

Lithologic Description and Areal Distribution. From Mansfield (1939: 8; condensed here): "The matrix of the Tamiami limestone [authors' note: the Ochopee Member] consists mainly of a dirty-white to gray, rather hard, porous, nonoölitic limestone with inclusions of clear quartz sand. The faunas, so far studied, include 6 genera of gastropods, 15 genera of pelecypods, and 2 genera of echinoids … Among the pelecypods, the scallops and oysters are the most conspicuous forms, both in numbers of species and individuals and in the rather large size that some of them attained. The echinoid *Encope macrophora tamiamiensis* Mansfield was found at three localities." Hunter's description of the Ochopee Limestone Member essentially follows Mansfield's: " … Light grey to white, hard, sandy limestone (a calcarenite) containing abundant identifiable mollusk molds and well preserved pectens, oysters, barnacles, and echinoids."

The Ochopee Member is confined to the southwestern Everglades area (Figure 4.8), in southern and western Collier, extreme southern Lee, northern Monroe, and western Dade Counties. The area corresponds to a large, shallow carbonate bank that was deposited along and behind the Immokalee Reef Tract, the southern end of the Immokalee Archipelago, and the southern part of the Hendry Platform. Carbonate fines eroding off this actively growing coral reef area accumulated in a large, fanlike structure and interfingered with the quartz sand banks of the central and northern Hendry Platform (the Pinecrest Member). Quartz sand also eroded off the Immokalee Archipelago and this, combined with the central Hendry Platform sands, contributed to the siliciclastic fraction seen in the typical Ochopee limestone. The "Bonita Springs Marl Member" of Missimer (1992) is here considered to be a geographically restricted, unconsolidated dolosilt facies of the Ochopee.

The Ochopee Member varies in thickness, from 10 m in northwestern Collier County, to over 20 m at Mule Pen Quarry, East Naples, Collier County, to over 15 m at the stratotype at Ochopee (Missimer, 1992). In the stratotype area of southeastern Collier, extreme northwestern Dade, and northernmost Monroe Counties, the Ochopee Member lies nearly at the surface, at 1–3 m depth. From there, the Ochopee dips to the west and south, occurring at depths below 10–20 m at the coast. These surface contours correspond to the geomorphology of the Pliocene carbonate bank. Several workers, such as Missimer (1992), have shown the Ochopee as being subjacent to the Pinecrest, and this has been adopted by many researchers and government offices. The presence of many key Pinecrest index fossils in the Ochopee (listed and illustrated in this section), however, readily demonstrates that the two members are contemporaneous, interfingering units. Workers such as Missimer, using well cores, have misidentified older Buckingham indurated beds as being the Ochopee, and this mistake has been carried on in many subsequent papers and technical reports. Because the member is so close to the surface near the stratotype area, the limestones were subject to extensive, long-term groundwater infiltration and leaching. Because of this, all aragonitic fossils have been dissolved away and remain only as casts and molds (often identifiable). Calcitic fossils, however, remained intact and are often very well preserved.

Stratotype. From Hunter (1968: 445; after Mansfield, 1939): " … rock pits [exist] between Carnestown (intersection of State Route 29 and U.S. Route 41) and the village of Ochopee, on U.S. Route 41, Collier County." As a point of interest, Ochopee became famous for having the smallest post office in the U.S. (a small booth).

Age and Correlative Units. Based on the presence of key stratigraphic index fossils, such as *Euvola hemicyclicus*, *Nodipecten collierensis*, and *Chesapecten madisonius*, the Ochopee Member is now known to date from the early middle Piacenzian Pliocene. The member is exactly correlative with the lower beds of the Mogarts Beach Member of the Yorktown Formation, Chesapeake Group of Virginia and northern North Carolina, the middle beds of the Duplin Formation of southern North Carolina and northern South Carolina, the upper beds of the Goose Creek Formation of South Carolina, and the lower beds of the *Cancellaria* Zone of the Jackson Bluff Formation of the Florida Panhandle.

Ochopee Index Fossils. Only calcitic fossils are preserved in the Ochopee, and these are mostly bivalves. Although containing some widespread mid-Piacenzian species, the Ochopee does contain a few species that have been found only in the stratotype area. This endemism demonstrates that the carbonate bank along the southern edge of the Hendry Platform must have had its own unique set of oceanographic parameters. This area supported immense aggregations of scallops belonging to several genera, including *Argopecten*, *Carolinapecten*, *Christinapecten*, *Chesapecten*, *Lindapecten*, *Nodipecten*, *Euvola*, and *Amusium*. The Ochopee Member is also the only Piacenzian-aged unit of the Everglades area to contain scallops of the genus *Chesapecten* (*C. madisonius*). The genus has never been found in any facies of the contemporaneous Pinecrest Member and must have been ecologically restricted to the southern Hendry Platform.

CHAPTER 6. The Tamiami Subsea (Everglades Pseudoatoll)

In the late 1970s, J.F. Meeder (then a graduate student at the University of Miami) discovered richly fossiliferous limestones and unconsolidated calcarenites that were being uncovered in housing developments, canal digs, and quarries in western Collier County. Upon closer examination, Meeder found that these beds represented zonated reef tracts and that they contained an unprecedented number of coral species, the richest ever found in the U.S. This important discovery was published in 1979, as an informal guide for a Miami Geological Society field trip to Collier County. In that work, Meeder illustrated several of the more prominent coral species, mapped the approximate extent of the reef systems, and discussed and illustrated the zonational patterns. A year later (1980), he published the first formal paper on the reef systems, describing the lithology of the limestones and associated carbonate facies, and discussing the reefal depositional environments. A complete survey of the reef assemblages of Collier and Lee Counties, including mapping, lithology, and faunal identification, was compiled by Meeder in his dissertation (1987, never published).

In 1984, while teaching at Florida International University in Miami, the senior author was notified of large local construction excavations that were bringing up beautifully preserved fossil corals and mollusks. These were prominent in spoil piles heaped around artificial lakes being dug for the Lakes of the Meadows housing development west of Miami on Bird Road. After several months of collecting and research, it became apparent that the Bird Road coral reef and molluscan assemblages were very similar to those found at several of Meeder's research sites in Collier County, such as at the Golden Gate housing development excavations and at the Mule Pen Quarry (later Florida Rock Industries, Inc., Naples Quarry). When comparing Meeder's discoveries in Collier County to the faunal and lithological data from the Bird Road reef sites, it was immediately apparent that the two faunas were the same; this uniformity of coral faunas was the first indication of the existence of a large-scale, previously-undiscovered state-wide carbonate feature. The senior author published these data the following year (Petuch, 1986), incorporating Meeder's research, along with that of the U.S. Geological Survey (Water Resources Division). In their study (Swayze and Miller, 1984), U.S. Geological Survey researchers showed that the reefs under Miami were actually part of a larger continuum of coral assemblages that extended all the way to northern Palm Beach County (the Zone of Secondary Permeability, a vuggy reefal limestone containing water), roughly aligning with the present-day Florida Turnpike. These fossil reef tracts, surrounding the southern Everglades in a "U"-shaped pattern (as described in Chapters 1 and 5), were referred to, collectively, as the **Everglades Pseudoatoll** (Petuch, 2004; Petuch and Roberts, 2007).

The immense Pliocene pseudoatoll, as described, was a geomorphologically-complex feature, with numerous sections and depositional environments (shown here on Figure 6.1). The three main coral reef tracts of this atoll-like structure included the **Immokalee Reef Tract** in the west and southwest, the **Miami Reef Tract** in the southeast, and the **Palm Beach Reef Tract** in the east, and these contained the largest number of species and genera of scleractinian corals found anywhere in the United States. Clusters of small coral bioherms and patch reefs, such as the **Long Pine Reefs** in the south, the **Okeechobee Reefs** in the north, and the **West Everglades Reefs** within the central lagoon contained less-diverse scleractinian faunas. These reef systems surrounded and enclosed a deep water lagoon, the **Loxahatchee Trough** and a shallow carbonate sediment shelf, the **Hendry Platform**. The wide, shallow **Kissimmee Embayment**, filled with mangrove forests and sea grass beds, connected to the northern end of the Loxahatchee Trough. The enclosed lagoon of the Tamiami Subsea had two main connections to open oceanic conditions; one in the northwest, the **Caloosahatchee Strait**, which connected to the Gulf of Mexico and one in the northeast, the **Loxahatchee Strait**, which connected to the Atlantic Ocean. Most of the water in the water in the deep Loxahatchee Trough flowed through these narrow, deep connections, constantly circulating during the tidal cycles.

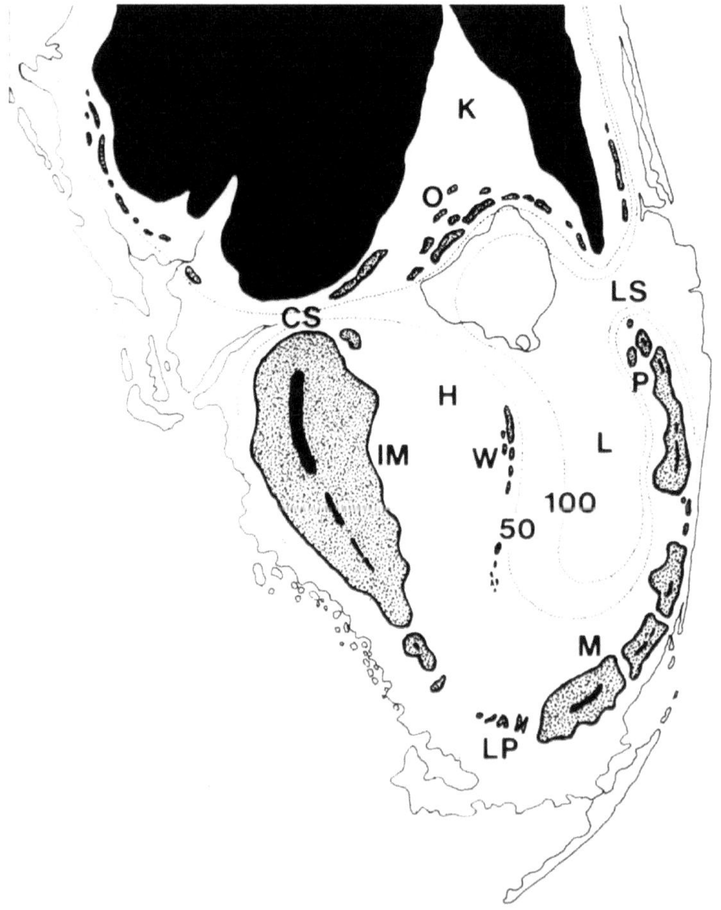

Figure 6.1 Map of the **Everglades Pseudoatoll** during the Pliocene, showing the main geomorphological features: **IM**= Immokalee Reef Tract; **LP**= Long Pine Reefs (coral bioherms); **M**= Miami Reef Tract; **P**= Palm Beach Reef Tract; **W**= West Everglades Reefs; **O**= Okeechobee Reefs (coral bioherms); **L**= Loxahatchee Trough; **H**= Hendry Platform; **K**= Kissimmee Embayment; **CS**= Caloosahatchee Strait; **LS**= Loxahatchee Strait. Contours are in meters below mean Pliocene sea level.

As described in Petuch (2004) and Petuch and Roberts (2007), the coral species of the Everglades Pseudoatoll were zonated by wave energy, being segregated into distinct, wide bands in which only a few species of corals dominated. Four broad, energy-related ecological zones are now known to have existed across most of the pseudoatoll reef tracts. These zones, defined by the amount of wave action, include (from open ocean towards shore): the **Reef Crest** (Zone 1, high-energy surf environment); the **Reef Platform** (Zone 2, lower energy surge environment, often subdivided into a front reef platform and a back reef platform); the **Back Reef** (Zone 3, low-energy environment); and the **Reef Lagoon** (Zone 4, quiet, deeper water environment). Collecting at several sites at the Bird Road lake excavations, on an east–west transect with each site separated by 0.5 km, the senior author (Petuch, 1986; 2004) was able to demonstrate the existence of this zonational pattern across the Miami Reef Tract section of the Everglades Pseudoatoll (shown here on Figure 6.2).

The Pliocene reefs at Mule Pen Quarry, the Golden Gate area near Naples, and at Bird Road contain the richest scleractinian coral fauna ever found in the Neogene western Atlantic, with over 100 species having been collected. The senior author (Petuch, 2004: 178) originally stated that the pseudoatoll reefs contained over 70 species, but subsequent analyses of Mule Pen Quarry material have shown that the number actually exceeds 100, and may reach 120. This species richness is unprecedented in the United States and approximates that of the present-day Moluccas Islands of Indonesia (the most diverse living coral fauna). Indeed, many of the coral genera found at Mule Pen are now extinct in the Atlantic Ocean but are flourishing in the South Pacific. At least half the pseudoatoll coral species were originally described from the Pliocene reefs of the Dominican Republic (Vaughan, 1919; Petuch, 2004: 176–185), demonstrating a strong Pliocene West Indian–Caribbean faunal influence. Some of the more abundant and frequently-collected scleractinians from the Pliocene Florida reef tracts include (arranged by the ecological zones on Figure 6.2):

Zone 1 Reef Crest: The highest-energy wave surge area of the reef tracts, housing the ***Pocillopora crassoramosa*** **Community** (Petuch, 2004:178), including corals such as:

Pocillopora crassoramosa (Figure 6.3A)
Pocillopora barracoaensis
Goniopora jacobiana (Figure 6.3B)
Astrocoenia meinzeri (Figure 6.3C)
Millepora ramosa (Figure 6.4E) (a hydrocoral, not a scleractinian)

Zone 2 Reef Platform: The moderate-energy area of the reef tracts, housing the ***Dichocoenia tuberosa*** **Community** (Petuch, 2004: 178-181), including corals such as:
Dichocoenia tuberosa (Figure 6.3D)
Dichocoenia cf. *eminans* (Figure 6.3F)
Dichocoenia species (Figure 6.3G)
Dendrogyra species (Figure 6.4G)
Barysmilia intermedia (Figure 6.4B)

Stephanocoenia spongiformis
Meandrina cf. *meandrites* (Figure 6.4F)
Montastrea brevis (Figure 6.3E)
Montastrea endothecata (Figure 6.4D)
Montastrea limbata
Solenastrea globosa (Figure 6.4C)
Siderastrea dalli
Siderastrea pliocenica (Figure 6.4A)

Zone 3 Back Reef: The low-energy protected area of the reef tracts, housing the ***Stylophora affinis*** **Community** (Petuch, 2004: 183), including corals such as:
Isophyllia desotoensis (Figure 6.5A)
Isophyllia cf. *sinuosa* (Figure 6.5E)
Mussa affinis (Figure 6.5B)
Mussa species (Figure 6.5C)
Syzygophyllia dentata (Figure 6.5F)
Asterosmilia exarata
Diploria sarasotana (Figure 6.5D)
Thysanus corbicula (Figure 6.5G)
Thysanus excentricus (Figure 6.5H)
Agaricia dominicensis (Figure 6.6A)
Stylophora affinis (Figure 6.6B)
Stylophora granulata
Stylophora minor (Figure 6.6E)
Porites barracoaensis
Porites matanzasensis
Acropora cf. *panamensis* (Figure 6.6C)
Acropora species (Figure 6.6D)
Oculina sarasotaensis

Zone 4 Reef Lagoon: quiet protected lagoon landward of the reef tracts; housing the ***Trachyphyllia bilobata*** **Community** (originally *Antillia bilobata*; Petuch, 2004: 183-185), including corals such as:
Septastrea matsoni (Figure 6.6I)
Antillia walli (Figure 6.6G)
Trachyphyllia bilobata (Figure 6.6F)
Manicina pliocenica (Figure 6.6H)
Placocyathus variabilis
Placocyathus barretti
Flabellum dubium
Flabellum exaratum

As can be seen from these scleractinian coral assemblages, the Reef Crest is dominated by low, massive, encrusting species. These share the shallow, high-energy habitat with coralline algae of the genera *Goniolithon* and *Lithothamnion*, and massive ramose forms of the hydrocoral *Millepora* (Petuch, 2004: plate 58). The seaward side of the Reef Platform is dominated by massive (sometimes over 2 m in diameter), rounded heads of various types of star corals. The back part of the Reef Platform, in quieter water conditions, is dominated by various types of mushroom, lettuce, and brain corals, primarily *Isophyllia*. The quiet, protected Back Reef is dominated by delicate branching corals such as *Stylophora* and *Acropora*, while the Reef Lagoon is dominated by solitary cup corals and rose corals. These lagoonal corals share the habitat with large beds of Turtle Grass (*Thalassia*) and open areas covered with beds of various types of large turritellid gastropods.

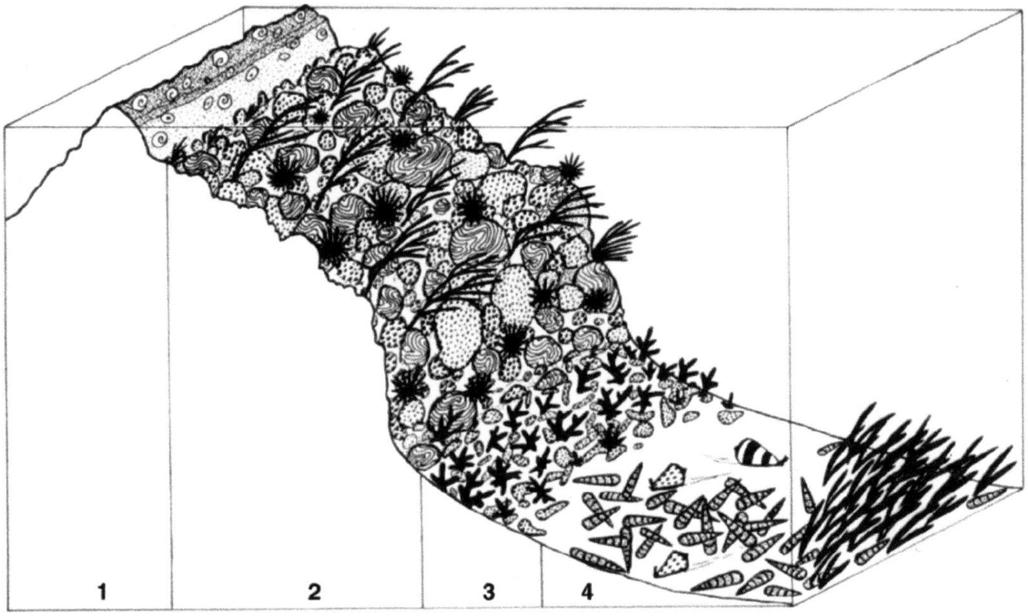

Figure 6.2 Schematic diagram showing a cross-section of a typical coral reef system of the Everglades Pseudoatoll during the late Pliocene. The ecological zones and their associated communities include: **Zone 1 (Reef Crest)**, the *Pocillopora crassoramosa* Community; **Zone 2 (Reef Platform)**, the *Dichocoenia tuberosa* Community; **Zone 3 (Back Reef)**, the *Stylophora affinis* Community; **Zone 4 (Reef Lagoon)**, the *Trachyphyllia bilobata* Community. (modified from Petuch, 2004; Petuch and Roberts, 2007).

Golden Gate Member, Tamiami Formation. In his overview of the Tamiami Formation, Missimer (1992: 67) resurrected the term "Golden Gate Reef" from Meeder's unpublished dissertation and elevated it to member status. We have shortened Missimer's informal name and referred to the reefal limestones that surround the Everglades basin as the Golden Gate Member. The reefal limestone and intercalated unconsolidated carbonate sediments of the member can be further subdivided into two main chronostratigraphic sections; the **lower Golden Gate** beds, which are contemporaneous with the Pinecrest Member, and the **upper Golden Gate** beds, which are contemporaneous with the Fruitville Member.

Lithologic Description and Areal Distribution. Building on the descriptions of Meeder and Missimer, Petuch and Roberts (2007:96) described the Golden Gate Member as "...an interbedded set of lithofacies dominated by limestones and unconsolidated calcilutites. These include vuggy moldic limestone with high primary porosity, highly fossiliferous limestone containing densely packed assemblages of well-preserved aragonitic corals (biomicrudites), layers of unconsolidated calcilutites and dolosilts packed with perfectly preserved mollusks and corals, and layers of spar-cemented recrystallized (calcitic) corals. The limestones vary in color from light gray to pale tan or cream. The unconsolidated calcilutites, dolosilts, and fine-grained

Figure 6.3 Corals of Zones 1 and 2, Everglades Pseudoatoll (Golden Gate Member).
A= *Pocillopora crassoramosa* Duncan, 1863, length 115 mm; **B**= *Goniopora jacobiana* Vaughan, 1919, length length 124 mm; **C**= *Astrocoenia meinzeri* Vaughan, 1919, length 55 mm; **D**= *Dichocoenia tuberosa* Duncan, 1863, length 63 mm; **E**= *Montastrea brevis* (Duncan, 1863), length 58 mm; **F**= *Dichocoenia* cf. *eminens* Weisbord, 1974, length 68 mm; **G**= *Dichocoenia* species, length 74 mm.

Figure 6.4 Corals of Zones 1 and 2, Everglades Pseudoatoll (Golden Gate Member).
A= *Siderastrea pliocenica* Vaughan, 1919, length 119 mm (showing an embedded specimen of the commensal muricid gastropod *Magilus streami*); **B=** *Barysmilia intermedia* Duncan, 1863, length 61mm; **C=** *Solenastrea globosa* Duncan, 1863, length 82 mm; **D=** *Montastrea endothecata* Duncan, 1863, length 128 mm; **E=** *Millepora ramosa* Esper, 1790, length 71 mm; **F=** *Meandrina* cf. *meandrites* (Linnaeus, 1767) (peduncular), length 205 mm; **G=** *Dendrogyra* species, length 43 mm.

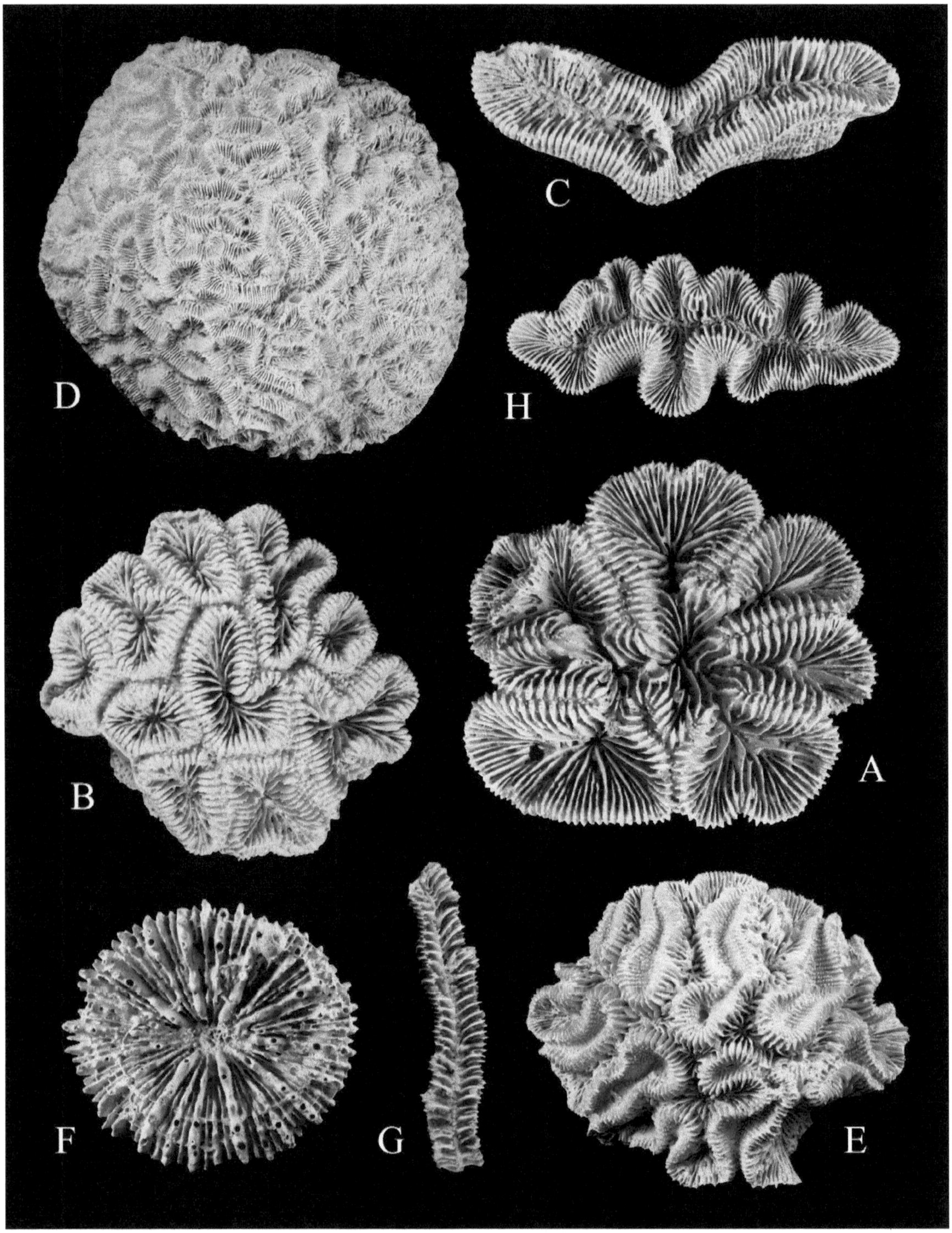

Figure 6.5 Corals of Zones 3 and 4, Everglades Pseudoatoll (Golden Gate Member).
A= *Isophyllia desotoensis* Weisbord, 1974, width 89 mm; **B**= *Mussa affinis* Duncan, 1863, width 87 mm; **C**= *Mussa* species, width 90 mm; **D**= *Diploria sarasotana* Weisbord, 1974, width 104 mm; **E**= *Isophyllia* cf. *sinuosa* (Ellis and Solander, 1786, width 87 mm; **F**= *Syzygophyllia dentata* (Duncan, 1863), width 43 mm; **G**= *Thysanus corbicula* Duncan, 1863, length 66 mm; **H**= *Thysanus excentricus* Duncan, 1863, length 70 mm.

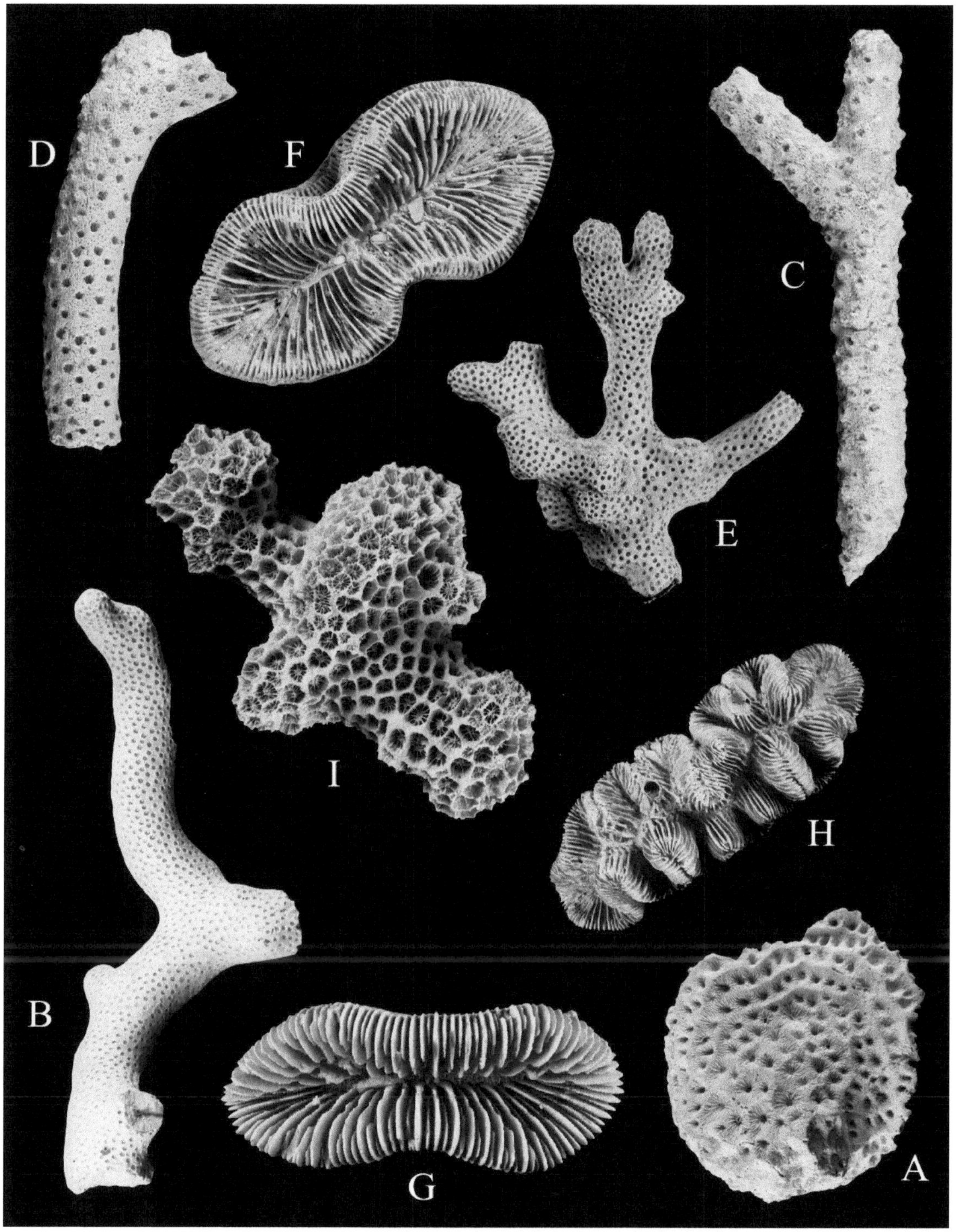

Figure 6.6 Corals of Zones 3 and 4, Everglades Pseudoatoll (Golden Gate Member).
A= *Agaricia dominicensis* Vaughan, 1919, width 54 mm; **B=** *Stylophora affinis* Duncan, 1863, length 121 mm; **C=** *Acropora* cf. *panamensis* Vaughan, 1919, length 74 mm; **D=** *Acropora* species, length 63 mm; **E=** *Stylophora minor* Duncan, 1863, length 65 mm; **F=** *Trachyphyllia bilobata* (Duncan, 1863), width 75 mm; **G=** *Antillia walli* (Duncan, 1863), length 47 mm; **H=** *Manicina pliocenica* Gane, 1900, length 85 mm; **I=** *Septastrea matsoni* Vaughan, 1919, length 68 mm.

calcarenites are colored pale tan, light gray, or pale blue-gray. Corals and mollusk shells preserved within these intercalated unconsolidated units are colored either pale tan or blue-gray, weathering to white on exposure. Small amounts of quartz sand are present in some of the limestones and calcilutites, rarely more than 15%. Some limestones, particularly those at the base of the member, are dolomitized and devoid of fossils. In some facies, thin stringers of quartz sand are present, usually at localities that were adjacent to Immokalee Island."

The areal distribution of the Golden Gate Member conforms to the shape of the reef tracts of the Everglades Pseudoatoll. The member is essentially "U"-shaped, following the outline of the main reef structures, from southern Lee, Collier, Monroe, Dade, Broward, and Palm Beach Counties. This reefal limestone member varies in thickness from 15 m at the Mule Pen Quarry, East Naples, Collier County, to 5 m in southern Collier County (Missimer, 1992: Core Log 11), to over 50 m under the Atlantic Coastal Ridge (Swayze and Miller, 1984). At the Mule Pen Quarry, the reefal limestones are 5 m below suface, whereas under eastern Palm Beach County (adjacent to the Florida Turnpike) they averaged 10 m below surface.

Stratotype. Wishing to preserve Meeder's original name (used extensively by Missimer), Petuch and Roberts (2007: 94-97) designated the type locality of the Golden Gate Member as the old Mule Pen Quarry (most recently, Florida Rock Industries, Inc., Naples Quarry) on Immokalee Road, East Naples, Collier County, near the Golden Gate housing subdivision (Sec. 23, T. 48 S., R. 26 E., Collier County). The quarry is now flooded as a large lake (over 3/4 mile across) surrounded by a housing development. The member is exposed at a depth of 5 m below the surface of the lake and can be accessed by scuba diving.

Age and Correlative Units. The basal beds of the Golden Gate Member, encountered at Bird Road, contained many classic Duplin, Yorktown, and Pinecrest index fossils, demonstrating that the reefal deposition started in the early middle Piacenzian Pliocene. Some of the widespread Duplin–Yorktown guide fossils found at Bird Road (Petuch, 1994) included the gastropods *Sconsia hodgei*, *Extractrix hoerlei*, *Ptychosalpinx multirugata*, *Contraconus adversarius*, and *Diodora (Glyphis) redemicula*. Only the upper beds of the Golden Gate were excavated during the last 10 years of excavations at the Mule Pen Quarry, Collier County, exposing the chronological equivalent of the Fruitville Member (described in the next chapter). Some of these younger index fossils, also found in the upper beds of the Yorktown and Duplin Formations, include the gastropod *Calliostoma conradianum* and the bivalve *Chama emmonsi*. The lower Golden Gate beds in Collier County, those equivalent to the Pinecrest and Ochopee Members, were shown by Missimer (1992) in his Core Log #11. The Golden Gate Member is now known to span the middle and late Piacenzian Pliocene, with the greatest development occurring during the late Piacenzian (Fruitville equivalent time). The member is exactly correlative with the

Mogarts Beach and Moore House Members of the Yorktown Formation and the Edenhouse Member of the Chowan River Formation, all of Virginia and northern North Carolina, the Duplin Formation of southern North Carolina and northern South Carolina, and the "*Cancellaria* Zone" of the Jackson Bluff Formation of the Florida Panhandle. The diachronous nature of Golden Gate deposition reflects the long-lived and continuous growth of the pseudoatoll coral reef systems.

Golden Gate Index Fossils. Besides the corals shown earlier in this chapter, the Phylum Mollusca also make up a large portion of the carbonate bioclast component, especially in the unconsolidated calcilutite and dolosilt beds. As in the corals, the molluscan fauna falls into two broad categories: a large component of endemic Floridian species and a smaller component of West Indian–Caribbean species, previously known only from the Pliocene of the Dominican Republic or Jamaica (Gurabo and Bowden Formations). Like the West Indian corals, these mollusks are only found on, or near, the pseudoatoll reefs and have not been found anywhere else in Florida. Some of these West Indian migrants (from the **Guraban Subprovince of the Gatunian Molluscan Province**; see Petuch, 2004) include the bursids *Marsupina proavus* and *Bursa amphitrites*, the muricids *Dermomurex olssoni* and *Coralliophila miocenica*, the ranellid *Septa henicum*, the cerithiids *Clypeomorus obesum* and *Cerithium turriculum*, the turbinid *Lithopoma aora*, the ampullariid *Pachychrommium guppyi,* and the xenophorid *Tugurium imperforatum*. Some of the more abundant Golden Gate index species are shown here in the following iconography.

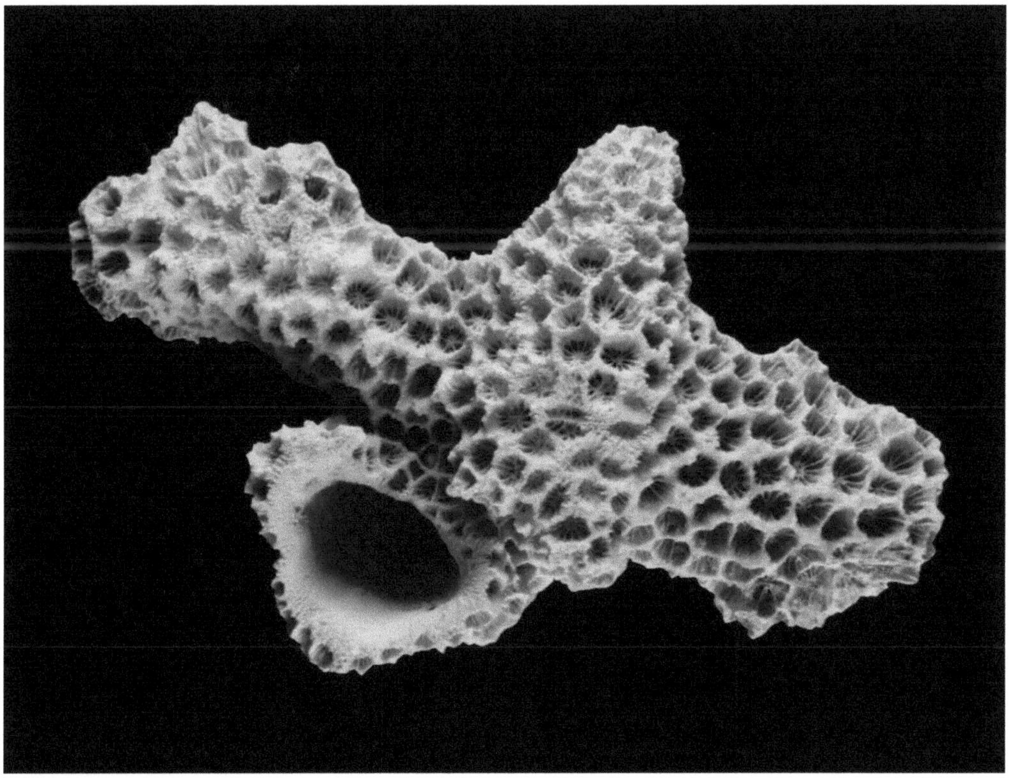

Septastrea matsoni Vaughan, 1919, length 68 mm, with the opening for a spiral chamber that was occupied by a commensal Hermit Crab.

Figure 6.7 Cowries of Zones 1 and 2, Everglades Pseudoatoll (Golden Gate Member).
A, B= *Pseudadusta lindae* (Petuch, 1986), length 47 mm; **C, D**= *Pseudadusta fehsei* Petuch and Drolshagen, 2011, length 65 mm; **E, F**= *Pseudadusta ketteri* (Petuch, 1994), length 54 mm; **G, H**= *Pseudadusta collierensis* Petuch and Drolshagen, 2011, length 42 mm; **I, J**= *Pseudadusta (Bithloa) irisae* Petuch and Drolshagen, 2011, length 59 mm; **K, L**= *Pseudadusta roachi* Petuch and Drolshagen, 2011, length 37 mm.

Figure 6.8 Mollusks of Zones 1 and 2, Everglades Pseudoatoll (Golden Gate Member). **A=** *Decoriatrivia miccosukee* Petuch, 1991, length 17 mm; **B=** *Pusula miamiensis* Petuch, 1991, length 18 mm; **C=** *Pusula dalli* Petuch, 1994, length 13 mm; **D=** *Cyphoma viaavensis* Petuch, 1986, length 26 mm; **E=** *Cyphoma finkli* Petuch, 1986, length 28 mm; **F=** *Cyphoma miamiensis* Petuch, 1986, length 21 mm; **G=** *Pseudocyphoma carolae* Petuch, 1986, length 20 mm; **H=** *Simnialena diegelae* Petuch, 1991, length 13 mm; **I=** *Calliostoma sincerum* Olsson and Harbison, 1953, length 26 mm; **J=** *Babelomurex lindae* Petuch, 1988, length 15 mm; **K=** *Babelomurex kissimmeensis* Petuch, 1994, length 28 mm; **L=** *Jenneria lindae* Petuch, 1988, length 31 mm (*J. violetae* is a synonym).

Figure 6.9 Mollusks of Zones 1 and 2, Everglades Pseudoatoll (Golden Gate Member).
A= *Subpterynotus miamiensis* Petuch, 1994, length 43 mm; **B=** *Murexiella stephensae* E. Vokes, 1994, length 19 mm; **C=** *Murexsul lyonsi* (Petuch, 1986), length 26 mm; **D=** *Chicoreus miamiensis* Petuch, 1994, length 74 mm; **E=** *Herndliconus irisae* (Petuch, 2004), length 33 mm; **F=** *Tequestaconus miamiensis* (Petuch, 1986), length 35 mm; **G=** *Tequestaconus antonioi* Petuch and Drolshagen, 2011, length 26 mm; **H=** *Magilus streami* Petuch, 1994, length 21 mm; **I=** *Aurantilaria lindae* (Petuch, 2004), length 76 mm (spire covered with coralline algae); **J=** *Polygona miamiensis* (Petuch, 1986), length 45 mm; **K=** *Parametaria lindae* Petuch, 1986, length 22 mm; **L=** *Bursa amphitrites* (Maury, 1917), length 16 mm (also found in the Gurabo Formation of the Domincan Republic).

Figure 6.10 Cowries of Zones 3 and 4, Everglades Pseudoatoll (Golden Gate Member).
A, B= *Akleistostoma (Mansfieldicypraea) miamiensis* Petuch and Drolshagen, 2011, length 62 mm; **C, D=** *Siphocypraea (Seminolecypraea) grovesi* Petuch, 2004, length 70 mm; **E, F=** *Siphocypraea (Lokossea) dimasi* Petuch, 1998, length 44 mm; **G, H=** *Siphocypraea (Lokossea) daughenbaughi* Berschauer and Waller, 2020, holotype, length 65.6 mm; **I, J=** *Siphocypraea (Lokossea) mulepenensis* Petuch, 1991, length 81 mm.

Figure 6.11 Mollusks of Zones 3 and 4, Everglades Pseudoatoll (Golden Gate Member).
A= *Aliger dominator* (Pilsbry and Johnson, 1917), length 231 mm; **B**= *Macrostrombus mulepenensis* (Petuch, 1994), length 170 mm; **C**= *Busycoarctum superbus* (Petuch, 1994), length 116 mm; **D**= *Lindafulgur miamiensis* (Petuch, 1991), length 152 mm; **E**= *Fulguropsis radula* Petuch, 1994, length 98 mm; **F**= *Macrostrombus collierensis* Petuch and Drolshagen, 2011, length 161 mm; **G**= *Titanostrombus immokaleensis* Petuch and Drolshagen, 2011, length 228 mm; **H**= *Busycon titan* Petuch, 1994, length 315 mm.

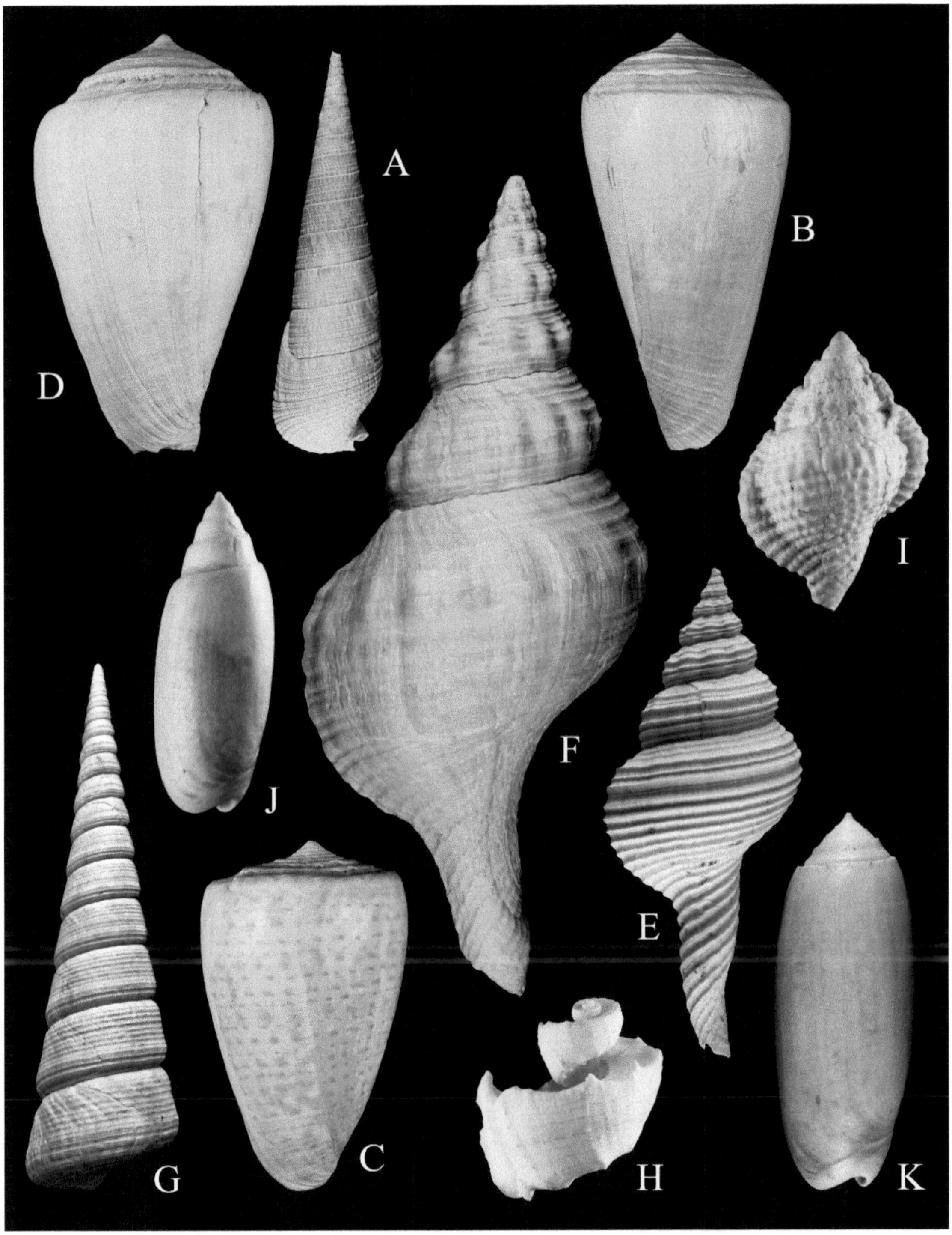

Figure 6.12 Mollusks of Zones 3 and 4, Everglades Pseudoatoll (Golden Gate Member).
A= *Cerithioclava dalli* (Olsson and Petit, 1964), length 75 mm; **B**= *Contraconus schmidti* (Petuch, 1991), length 98 mm; **C**= *Lindaconus martinshugari* (Petuch, 1994), length 83 mm; **D**= *Pyruconus leonensis* (Mansfield, 1930), 142 mm; **E**= *Heilprinia miamiensis* Petuch, 1994, length 74 mm; **F**= *Triplofusus harveyensis* (Mansfield, 1930), length 287 mm; **G**= *Eichwaldiella pontoni* (Mansfield, 1931), length 165 mm; **H**= *Extractrix hoerlei* (Olsson, 1967), length 18 mm **I**= *Marsupina proavus* (Pilsbry, 1922), length 22 mm (also found in the Gurabo Formation of the Dominican Republic); **J**= *Americoliva susanae* (Petuch, 1994), length 37 mm; **K**= *Americoliva mansfieldi* (Petuch, 1994), length 65 mm.

View of an excavation in the Bonita Grande Aggregates quarry, Bonita Springs, Lee County, Florida. The shattered rocks that are piled up in the foreground, and in the large piles next to the excavator crane, are composed of coral reef limestone from the Golden Gate Member of the Tamiami Formation. Over 100 species of Pliocene corals have been collected here, along with several hundred species of reef-dwelling mollusks. Photo courtesy of Melanie Briskin.

CHAPTER 7. The Tamiami Subsea (Tamiami Formation: Fruitville Member)

Taking into account the lack of formal descriptions for most of the Tamiami members, Waldrop and Wilson (1990) proposed a new formational name for the Pliocene sequence at Sarasota; the "Fruitville Formation" (named for Fruitville, a small town east of the city of Sarasota). All the shell beds above Unit 11 at the APAC pit (Figure 5.3) were included in their new formation, and the name was presented as a replacement for the loosely described "Pinecrest Beds" of Olsson. Upon closer scrutiny, it can be seen that Waldrop and Wilson were utilizing only biostratigraphic analyses, in effect describing a biozone and not a true lithostratigraphic formation. Because of the lack of a detailed lithologic description, Petuch and Roberts (2007:104) recommended that Waldron and Wilson's "Fruitville Formation" not be afforded full formational status. In light of better lithologic descriptions of the Pinecrest Member (Missimer, 1992; earlier in this chapter), Petuch and Roberts removed Units 5–9 of the APAC section from Waldrop and Wilson's "Fruitville Formation" and included these units in an expanded definition of the Tamiami Formation. Because several workers (such as Emily Vokes, in her later muricid studies in *Tulane Studies*) have used the name "Fruitville Formation" for the Sarasota beds, we feel it is best to preserve and redefine the name. In this book, as in Petuch and Roberts (2007), we restrict Waldrop and Wilson's name to the upper Tamiami beds, corresponding to Unit 2, Unit 3, and Unit 4 in the APAC pit (Figure 5.3). These are referred to here as the **Fruitville Member**.

Lithologic Description of the Fruitville Member. The Fruitville Member was deposited in two distinct pulses, corresponding to two eustatic highs during the late Pliocene (Figure 5.2). Each of these marine transgressions were separated by at least 200,000 years of dry, terrestrial environments, producing obvious unconformities. This sequence is best seen in the APAC pit at Sarasota (Figure 5.3), which can be used as a standard reference section showing subjacent and suprajacent units. The first transgression, at 3 million years B.P., had the highest sea level and deposited Unit 3 and Unit 4 and their equivalents around the state. The second transgression, at 2.6 million years B.P., had a lower sea level and deposited Unit 2 and its equivalents. When discussing the Fruitville Member in this book, we will use these numbered units to differentiate positions within the depositional sequence.

Based on Waldron and Wilson's paper, Petuch and Roberts (2007:105-106) redefined the Fruitville Member as "…a set of interbedded units variably composed of organic-rich quartz sand, muddy quartz sand with densely packed mollusk shell bioclasts (usually bivalves), and quartz sand with molluscan bioclasts and small amounts of clay, usually sepiolite. Limestones are absent from the Fruitville Member, although the bedding plain surfaces of some units, such as Unit 3 and Unit 4 at Sarasota, have thin (less than 10 cm) indurated duricrusts composed of fossiliferous arenaceous calcarenites. The quartz sand varies in color from white, to yellow-tan, to dark brown or dark gray. In the organic-rich sand units (such as Unit 4 at Sarasota),

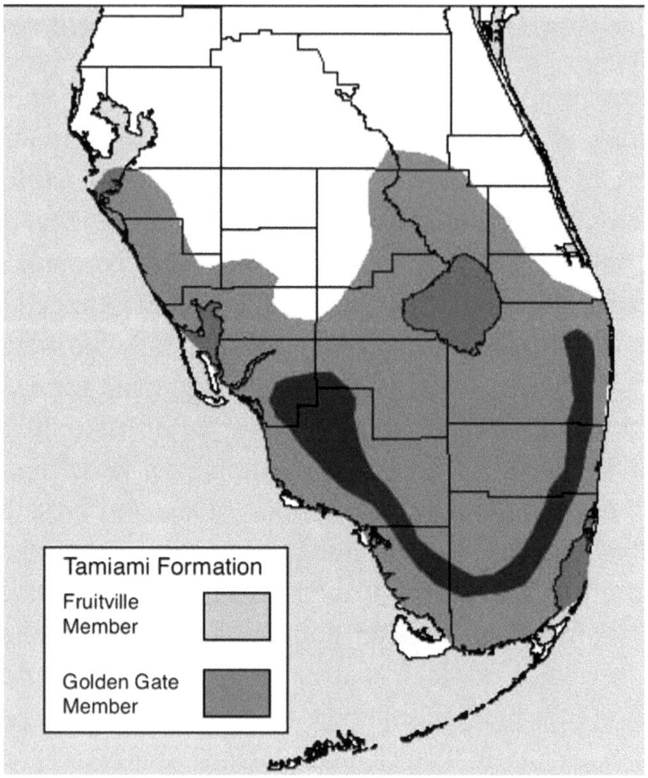

Figure 7.1 Areal Distribution of the Fruitville and Golden Gate (upper beds) Members of the Tamiami Formation in southern Florida. The coral reef-associated index fossils for the upper beds of the Golden Gate Member are discussed and illustrated in Chapter 4.

the beds are filled with black-stained mollusks and have a high sulfide content, with small crystals of pyrite being deposited on the shells and quartz sand grains. Fruitville molluscan fossils are very well preserved and range in color from white, to light brown, to bluish-black. These shells make up the largest portion of the carbonate fraction. The Fruitville can be differentiated from the underlying Pinecrest by having thinner quartz sand units that are muddier and contain more clay and by having beds that often contain high concentrations of pyrite crystals."

The areal distribution of the Fruitville Member (Figure 7.1) is essentially the same as the underlying Pinecrest Member. Like the Pinecrest, the Fruitville also has a distinctive Kissimmee Facies, composed of muddy, organic-rich sand generally packed with crushed mollusk shells ("shell hash"). The Kissimmee Embayment during Fruitville time also housed a molluscan fauna that was rich in endemic species. The member varies in thickness from 5 m at Sarasota, Sarasota County, to 8 m in Lee County, to 7 m at Fort Drum, Okeechobee County (Rucks Pit). The surface contours of the Fruitville Member conform to the bathymetric contours of the Tamiami Subsea. At Sarasota, the Fruitville lies 2–3 m below land surface, whereas at Fort Drum the member lies approximately 10–12 m below land surface.

Stratotype. We here designate the section in the APAC pit, Unit 2, Unit 3, and Unit 4 (Figure 4.3), as the stratotype of the Fruitville Member. The APAC quarry is now a lake in a Sarasota County park between Richardson Road and Interstate Highway 75 (I-75), in the Fruitville area of Sarasota, Sarasota County (1 km north of Fruitville Road, Sec. 12, T. 36 S., R. 18 E.). The member lies approximately 2 m below the lake surface and can be accessed by scuba diving.

Age and Correlative Units. Based on the presence of widespread molluscan chronostratigraphic index fossils such as the bivalves *Carolinapecten bertiensis*, *Stralopecten ernsestmithi*, *Nodipecten vaccamavensis*, and *Chama emmonsi* and the gastropods *Monostiolum petiti* and *Sinistrofulgur adversarium*, the Fruitville Member is now known to date from the late Piacenzian Pliocene. The member is correlative with the Moore House Member of the Yorktown Formation (Unit 3 and Unit 4 only) and the Edenhouse Member of the Chowan River Formation (Unit 2 only), both of Virginia and northern North Carolina, the lower beds of the Waccamaw Formation of southern North Carolina and northern South Carolina, the uppermost beds of the "*Cancellaria Zone*" of the Jackson Bluff Formation of the Florida Panhandle, and the upper beds of the Golden Gate Member.

Fruitville Member Unit 4. The basal bed of the Fruitville Member, Unit 4 of the APAC type section in Sarasota, is the thickest and best-developed of the three Fruitville beds. At that locality, the thick Unit 4 bed was one of the most conspicuous stratigraphic units, containing dense beds of fossil mollusks, all stained a distinctive blue-black color from having been deposited within a reducing, organic-rich environment. This manganese sulfide and iron sulfide staining is so pervasive within the unit that the senior author, in the original stratigraphic assessment of the Pliocene series in Sarasota (Petuch, 1982; 1994), referred to this distinctively-colored bed as the "Black Layer" and used it as a reference marker bed for determining the sequence stratigraphy of the suprajacent and subjacent strata. At the type section in the APAC pit, Unit 4 contained a recognizable brackish water estuarine fauna, being dominated by the estuarine mactrid bivalve, *Mulinia sapotilla* (Figures 7.2 and 7.7F), and by abundant potamidid gastropods such as *Pyrazisinus scalinus*, *P. lindae*, *P. sarasotaensis*, *Cerithidea lindae*, (all shown here on Figure 7.7), and *Potamides matsoni*, and the melongenid *Melongena draperi* (the paleoecology of the *Mulinia sapotilla* Community was described by the senior author; Petuch, 2004:187). The estuarine reduced sediments and shell beds of Unit 4 at the APAC Pit are typical of those deposited within the extensive **Myakka Lagoon System**, an immense deltaic area at the mouths of the proto-Myakka, proto-Manatee, and proto-Braden Rivers. The Myakka Lagoon System (named for the Myakka River; Petuch, 2004; Petuch and Roberts, 2007), extended from what is now southern Tampa Bay, Hillsborough County south to the present-day Peace River, Charlotte County and extended inland as far as eastern Sarasota County. This is the largest and best-developed paleo-estuary known from anywhere in Florida.

During the highest sea level stand of the late Piacenzian (Unit 4 time), the Myakka Lagoon System was flooded to its maximum size and contained the largest and best-developed mangrove forests of the entire Tamiami Subsea region. These extensive estuarine forests and intertidal mud flats, which most probably resembled the mangrove jungles found along present-day southern New Guinea, housed the richest estuarine molluscan fauna known from the American Pliocene. Large pieces of carbonized mangrove wood, probably from several tree genera and species, are regularly emcountered mixed together with the dense beds of *Mulinia* clams, demonstrating that the mangrove jungles grew in close proximity to the tidal channels and intertidal mud flats. These areas supported a rich fauna of cowrie shells, with at least 9 endemic species known to date (Petuch, Berschauer, and Myers, 2018; shown here on Figures 7.4 and 7.5). Of these, the neotenic genus *Calusacypraea* and its subgenus *Myakkacypraea* are known only from the Myakka Lagoon area, where they probably lived a semi-amphibious lifestyle, crawling on, and feeding among, the dense tangles of mangrove pneumatophores (respiratory roots) that were exposed at low tide. These highly-restricted endemic cowrie genera have been found nowhere else in Florida. Deeper tidal channels between the intertidal mud flats were subject to tidal flushing, with the currents scouring the channel floors, removing organic material and winnowing the loose shells on the sea floor (shown here on Figure 7.2). These cleaner sand areas appear to have supported sea grass beds and much more species-rich molluscan assemblages (see Figure 7.3).

Other Unit 4 stratigraphic equivalents are known from around the Everglades region, with the best-known being the lower Fruitville beds exposed in excavations and quarries around the Kissimmee Valley and Kissimmee River. During the Pliocene, this area formed the axis of the elongated **Kissimmee Embayment**, a narrow bay that extended into central Florida, well northof the main Tamiami Subsea (see Figure 6.1, Chapter 6). Like the Myakka Lagoon System, the Kissimmee Embayment contained extensive mangrove forests and intertidal sand flats, but because of tidal flushing from the Loxahatchee Trough area to the south and the lack of major fresh water input from rivers, it contained far fewer organic-rich reducing environments. The cleaner sedimentary regimes of the Kissimmee Embayment contained a highly endemic molluscan fauna, with a very different composition from that seen in the Myakka estuaries. The majority of the characteristic index species found within the Kissimmee area were restricted to the central and northern areas, occurring exclusively on the extensive clean carbonate sediment shoals that formed within the quiet, sheltered waters of the embayment. The intervening Polk Peninsula and the deep Caloosahatchee Strait acted as ecological barriers, preventing mollusks from the Myakka Lagoon System from migrating into the Kissimmee Embayment. Because of these barriers during Unit 4 time, the northern sections of the Tamiami Subsea contained two distinct and separate molluscan faunas. The characteristic Unit 4 index taxa, from both the Myakka Lagoon System and the Kissimmee Embayment, are shown here in the following iconography.

Figure 7.2 View of the basal beds of the Fruitville Member (Unit 4) of the Tamiami Formation in the APAC Pit, Sarasota, Sarasota County, Florida. Here, closely packed beds of the small bivalve *Mulinia sapotilla* make up almost the entire unit. Thin layers of the large pearly mussel *Perna conradiana* are intercalated between the *Mulinia sapotilla* beds and these were deposited through winnowing by tidal currents. In the lower left corner, numerous specimens of *Melongena draperi* have accumulated within a lag deposit. This is a classic **Myakka Lagoon** ecosystem.

Figure 7.3 Close-up of an exposure of the lower Fruitville Member (Unit 4), Tamiami Formation, in the SMR Pit, Sarasota, Florida, showing the *in situ* specimens of the cowrie *Akleistostoma (Mansfieldicypraea) macbrideae*, the dwarf whelk *Pyruella schmidti*, and numerous bivalve fragments. This assemblage was deposited within a tidal channel that had formed by tidal currents flowing between shallow mud flats. Photo by Anton Oleinik.

Fruitville Member Unit 3. The middle bed of the Fruitville Member, Unit 3 of the APAC type section at Sarasota, is the second-thickest of the Fruitville beds, but is generally thinner and less-developed than the Unit 4 beds. This implies a shorter transgressional period of deposition and also a larger erosional unconformity that resulted from a longer period of subsequent regression. By Unit 3 time at Sarasota, the sediments deposited in the Myakka Lagoon System were noticeably cleaner, containing a far smaller percentage of organics and with no sign of reducing environments. This implies that there was less fresh water input from river systems and that most of the mangrove jungles were diminished in size and were no longer present in the areas of the APAC and Quality Aggregates Pits. Instead, the Sarasota area in Unit 3 time was characterized by large, shallow shoals within a quiet, open sound and with normal salinity water conditions (35 parts per thousand). These protected, cleaner sand areas, supported extensive beds of the large pearly mussel, *Perna conradiana* (Figures 7.10, 7.11, and 7.14G), and these formed the basis for a rich and highly endemic ecosystem (the ***Perna conradiana*** **Community**; see Petuch, 2004: 187).

Like the subjacent Unit 4, the Unit 3 beds at Sarasota have equivalent strata in the Kissimmee Embayment area, most frequently encountered along the Kissimmee River (during channelizing excavations) and at the Rucks Pit in Fort Drum, Okeechobee

Figure 7.4 Estuarine Cowries of the lower Fruitville Member (Unit 4), Tamiami Formation. **A, B**= *Akleistostoma (Akleistostoma) olssoni* Petuch and Drolshagen, 2011, length 82 mm; **C, D**= *Akleistostoma (Mansfieldicypraea) macbrideae* (Petuch, 1998); length 78 mm; **E, F**= *Akleistostoma (Paludacypraea) cookei* Petuch and Frolshagen, 2011, length 55 mm; **G, H**= *Akleistostoma (Paludacypraea) fruitvillensis* Petuch and Drolshagen, 2011, length 58 mm; **I, J**= *Akleistostoma (Dallicypraea) williamdalli* Petuch and Drolshagen, 2011, length 73 mm. All of these species were restricted to the Myakka Lagoon System.

Figure 7.5 Estuarine Cowries of the lower Fruitville Member (Unit 4), Tamiami Formation. **A, B=** *Pahayokea (Gardnericypraea) jenniferae* (Petuch, 1998), length 81 mm; **C, D=** *Calusacypraea tequesta* (Petuch, 1996), length 75 mm; **E, F=** *Calusacypraea (Myakkacypraea) myakka* Petuch, 2004, length 59 mm; **G, H=** *Pseudadusta metae* (Petuch, 1994), length 63 mm. All of these species were restricted to the Myakka Lagoon System.

Figure 7.6 Estuarine Gastropods of the lower Fruitville Member (Unit 4), Tamiami Formation. **A**= *Tropochasca metae* Petuch, 1994, length 94 mm; **B**= *Echinofulgur jonesae* Petuch, 1994, length 75 mm; **C**= *Melongena draperi* Petuch, 2004, length 65 mm; **D**= *Pyruella schmidti* Petuch, 1994, length 55 mm; **E**= *Laevisycon laevis* (Petuch, 1982), length 64 mm; **F**= *Hystrivasum lindae* Petuch, 1994, length 84 mm; **G**= *Pleioptygma nancyae* Petuch, 1994, length 94 mm; **H**= *Strombus sarasotaensis* Petuch and Drolshagen, 2011, length 97 mm.

Figure 7.7 Estuarine Mollusks of the lower Fruitville Member (Unit 4), Tamiami Formation. **A**= *Pyrazisinus sarasotaensis* Petuch, 1994, length 31 mm; **B**= *Pyrazisinus lindae* Petuch, 1994, length 53 mm; **C**= *Pyrazisinus scalinus* (Olsson, 1967), length 48 mm; **D**= *Cerithidea lindae* Petuch, 1994, length 29 mm; **E**= *Bulla sarasotaensis* Petuch, 1994, length 20 mm; **F**= *Mulinia sapotilla* Dall, 1898, length 28 mm; **G**= *Stramonita sarasotaensis* Petuch, 1994, length 30 mm; **H**= *Contraconus berryi* Petuch, 1994, length 78 mm; **I**= *Scaphella petiti* Petuch, 1994, length 95 mm.

Figure 7.8 Cowries of the lower Fruitville Member (Unit 4, Kissimmee Facies), Tamiami Formation.

A, B= *Akleistostoma (Ingramicypraea) cliffordi* Petuch and Drolshagen, 2011, length 47 mm; **B, C=** *Akleistostoma (Olssonicypraea) bairdi* Petuch, 2004, length 44 mm; **D, E=** *Pahayokea rucksorum* (Petuch, 2004), length 69 mm; **F, G=** *Pahayokea gabrielleae* (Petuch, 2004), length 69 mm; **I, J=** *Akleistostoma (Paludacypraea) judei* (Petuch and Drolshagen, 2011), length 44 mm. All of these species were restricted to the Kissimmee Embayment.

Figure 7.9 Gastropods of the lower Fruitville Member (Unit 4, Kissimmee Facies), Tamiami Formation.

A, B= *Pahayokea (Pahayokea) basingerensis* (Petuch, 2004), length 62 mm; **C, D**= *Pahayokea (Kissimmecypraea) eddiematchetti* Petuch and Drolshagen, 2011, length 79 mm; **E, F**= *Okeechobea waldroni* Petuch and Drolshagen, 2011, length 57 mm; **G, H**= *Okeechobea (Yeehawcypraea) matchetti* Petuch and Drolshagen, 2011, length 58 mm; **I**= *Pyruella basingerensis* Petuch, 1994, length 92 mm; **J**= *Hystrivasum kissimmense* (Hollister, 1971), length 135 mm. All of these species were restricted to the Kissimmee Embayment.

County. These Kissimmee Embayment beds are lithologically-similar to Unit 7 and Unit 4 equivalent sediments, implying that the local depositional environments changed little over the entire late Pliocene, retaining a clean, open-oceanic water chemistry and a low percentage of organics. The extensive beds of the mussel *Perna conradiana*, which dominated the Myakka Lagoon area during Unit 3 time, are completely absent from stratigraphic units that were deposited within the Kissimmee Embayment. These characteristic Unit 3 Myakka Lagoon mussel beds are replaced, within the Kissimmee Embayment area, by molluscan assembages that are indicative of sea grass (probably Turtle Grass, *Thalassia*) and open sea floor environments. The molluscan fauna of the Kissimmee region, like that of Unit 4, contains a high percentage of endemic, highly-restricted species, demonstrating that there was little or no genetic interchange between the Myakka Lagoon System and the Kissimmee Embayment during Unit 3 time. Some of the more important and characteristic Unit 3 index fossils are shown here in the following iconography.

Fruitville Member Unit 2. The thinnest bed of the Fruitville Member, Unit 2 of the APAC type section at Sarasota, represents the final marine transgression of the Pliocene. Around southern Florida, Unit 2 has only been recognized at a few localities and probably is represented by scattered erosional stringers that survived the massive dissolution during the sea level drops at the Plio-Pleistocene boundary (Piacenzian-Gelasian boundary). The best-developed stringers of Unit 2 have been recognized from the APAC and Quality Aggregates quarries at Sarasota and in the Rucks Pit at Fort Drum, Okeechobee County. At Sarasota, the Unit 2 beds contained closely-packed assemblages of bivalves, especially spiny *Arcinella* cf. *cornuta*, large venerids such as *Mercenaria* cf. *carolinensis* and *Dosinia* species, *Laevicardium* cardiids, and *Atrina* pen shells (Figure 7.18). At Fort Drum, Okeechobee County, the Unit 2 beds contained closely-packed beds of the spiny slipper shell, *Crepidula chamnessi*, which apparently lived attached to Turtle Grass. All of these units and ecosystems housed highly endemic molluscan faunas, and these can be used as stratigraphic guide fossils. During Unit 2 time, the northern end of the Kissimmee Embayment was connected to the Atlantic Ocean by a series of narrow cuts across the St. Lucie Peninsula. This allowed cooler water conditions to enter parts of the embayment, producing marine conditions more similar to the Carolinas than Florida. This cooler marine climate also allowed a large part of the Chowan River Formation (Edenhouse Member) molluscan fauna to invade the Okeechobean Sea. These Virginia and Carolina species remained in the northern end of the Kissimmee Embayment and moved no farther south. Unit 2 also contains the last known species of several prominent Pliocene cowrie groups, such as the genera *Akleistostoma*, *Calusacypraea*, and *Pseudadusta* and the subgenera *Mansfieldicypraea*, *Seminolecypraea*, *Myakkacypraea,* and *Bithloa* from the Myakka Lagoon area and the subgenera *Ingramicypraea*, *Olssonicypraea*, *Yeehawcypraea*, and *Kissimmecypraea* from the Kissimmee Embayment area. All of these important Pliocene Okeechobean cowrie taxa, along with the spiny whelk *Tropochasca* and the cone shell genus *Fenestraconus*, became extinct, abruptly, at the end of Unit 2 time, during the climatic degeneration and sea level drops at the Plio-Pleistocene Boundary.

Figure 7.10 View of an exposure of the upper beds of the Fruitville Member, Tamiami Formation in Quality Aggregates Pit #6, Sarasota, Sarasota County, Florida. The base of the road is the bedding plane surface of Unit 4, while Unit 3 has been cut back to produce a line of low cliffs. Unit 2, the uppermost Fruitville bed, is the darker-colored layer higher up on the slope, below the trees and bushes. Photo by Brian Schnirel.

Figure 7.11 Close-up of the exposed surface of Unit 3 of the Fruitville Member, Tamiami Formation in Quality Aggregates Pit #6, Sarasota, Sarasota County, Florida. The large cowrie, *Siphocypraea cannoni*, is exposed (to the left of the hammer) along with numerous specimens of the crepidulid gastropods *Trochita floridana* and the triangular *Crepidula cannoni*, and two specimens of the giant miter shell *Pleioptygma saginata* (in the upper left corner). These gastropods were buried in a matrix of crushed *Perna conradiana* pearly mussel shells.

Figure 7.12 Cowrie Shells of the middle Fruitville Member (Unit 3), Tamiami Formation.
A, B= *Akleistostoma (Akleistostoma) rilkoi* (Petuch, 1998), length 66mm; **C, D=** *Akleistostoma (Mansfieldicypraea) juliagardnerae* Petuch and Drolshagen, 2011, length 78 mm; **E, F=** *Akleistostoma (Paludacypraea) walleri* Petuch, Berschauer, and Myers, 2018, length 68 mm; **G, H=** *Siphocypraea cannoni* Petuch, 1994, length 74 mm; **I, J=** *Siphocypraea (Seminolecypraea) parodizi* Petuch, 1994, length 57 mm. All of these species were endemic to the Myakka Lagoon System.

Figure 7.13 Cowrie Shells of the middle Fruitville Member (Unit 3), Tamiami Formation.
A, B= *Pahayokea (Gardnericypraea) buchecki* Petuch and Drolshagen, 2011, length 80 mm; **C, D**= *Calusacypraea (Calusacypraea) sarasotaensis* (Petuch, 1994), length 41 mm; **E, F**= *Calusacypraea (Myakkacypraea) kelleyi* (Petuch, 1998), length 38 mm; **G, H**= *Pseudadusta (Bithloa) kalafuti* (Petuch, 1994), length 84 mm. All of these species were endemic to the Myakka Lagoon System during Unit 3 time.

Figure 7.14 Mollusks of the middle Fruitville Member (Unit 3), Tamiami Formation.
A= *Pyruella federicoae* Petuch, 1994, length 78 mm; **B**= *Laevisycon turbinalis* (Petuch, 1982), length 55 mm; **C**= *Pterorhytis roxaneae* Petuch, 1994, length 38 mm; **D**= *Pterorhytis squamulosa* Petuch, 1994, length 36 mm; **E**= *Echinofulgur helenae* Olsson, 1967, length 98 mm; **F**= *Hexaplex jameshoubricki* Petuch, 1994, length 113 mm; **G**= *Perna conradiana* (d'Orbigny, 1852), length 109 mm; **H, I**= *Cassis ketteri* Parodiz and Tripp, 1992, length 154 mm.

Figure 7.15 Gastropods of the middle Fruitville Member (Unit 3), Tamiami Formation.
A= *Chicoreus miccosukee* Petuch, 1994, length 35 mm; **B**= *Scaphella gravesae* Petuch, 1994, length 153 mm; **C**= *Scaphella maureenae* Petuch, 1994, length 59 mm; **D**= *Hystrivasum chilesi* Petuch, 1994, length 105 mm; **E**= *Hystrivasum jacksonense* (E. Vokes, 1966), length 119 mm; **F**= *Melongena cannoni* Petuch, 1994, length 137 mm; **G**= *Strombus cannoni* Petuch and Drolshagen, 2011, length 81 mm; **H**= *Ximeniconus marylandicus* (Green, 1830), length 30 mm. **I**= *Lindaconus streami* (Petuch, 1994), length 62 mm.

Figure 7.16 Gastropods of the Kissimmee Facies of the middle Fruitville Member (Unit 3), Tamiami Formation.
A, B= *Akleistostoma (Olssonicypraea) diegelae* (Petuch, 1994), length 29 mm; **C, D=** *Pahayokea penningtonorum* (Petuch, 1994), length 50 mm; **E, F=** *Pahayokea heimeri* Petuch and Drolshagen, 2011, length 56 mm; **G, H=** *Pahayokea (Kissimmecypraea) leonardi* Petuch and Drolshagen, 2011, length 68 mm; **I=** *Hystrivasum palmerae* (Hollister, 1971), length 81 mm; **J=** *Hystrivasum squamosum* (Hollister, 1971), length 125 mm. All of these species were endemic to the Kissimmee Embayment.

Figure 7.17 Gastropods of the Kissimmee Facies of the middle Fruitville Member (Unit 3), Tamiami Formation.

A= *Tropochasca kissimmeensis* Petuch, 1994, length 73 mm; **B=** *Pliculofusus diegelae* (Petuch, 1994), length 95 mm; **C=** *Pliculofusus maryae* (Petuch, 1994), length 149 mm; **D=** *Pyrazisinus kissimmeensis* (Olsson, 1967), length 82 mm; **E=** *Heilprinia florida* (Olsson and Harbison, 1953), length 118 mm; **F=** *Eupleura kissimmeensis* Petuch, 1994, length 24 mm; **G=** *Hystrivasum shrinerae* (Hollister, 1971), length 102 mm; **H=** *Pleioptygma kissimmeensis* Petuch, 1994, length 94 mm; **I=** *Pliculofusus okeechobeensis* (Petuch, 1994), length 94 mm. All of these species were endemic to the Kissimmee Embayment.

Figure 7.18 View of an exposure of the upper bed of the Fruitville Member (Unit 2), Tamiami Formation, in Quality Aggregates Pit #9, University Drive, Sarasota, Sarasota County, Florida. A pair of the large venerid bivalve, *Mercenaria* cf. *carolinensis*, is exposed at the center of the photograph and several specimens of the spiny jewel box clam, *Arcinella* cf. *cornuta*, are exposed along the bottom-right. The exposure is stained red from iron oxide.

Figure 7.19 Cowries of the upper Fruitville Member (Unit 2), Tamiami Formation.
A, B= *Akleistostoma adrienae* Petuch and Drolshagen, 2011, length 60 mm; **C, D**= *Akleistostoma (Mansfieldicypraea) ivyi* Petuch and Drolshagen, 2011, length 62 mm; **E, F** = *Siphocypraea janowskyi* Petuch and Drolshagen, 2011, length 77 mm; **G, H** = *Siphocypraea (Seminolecypraea) alligator* Petuch, 1994, length 64 mm; **I, J** = *Calusacypraea loxa* Petuch and Drolshagen, 2011, length 42 mm.

Figure 7.20 Cowries of the upper Fruitville Member (Unit 2), Tamiami Formation.
A, B= *Calusacypraea (Myakkacypraea) schnireli* Petuch, 2004, length 41 mm; **C, D**= *Pseudadusta marilynae* (Petuch, 1994), 42 mm; **E, F**= *Pseudadusta yingsti* Petuch and Drolshagen, 2011, length 41 mm; **G, H**= *Pseudadusta (Bithloa) juyingae* Petuch and Drolshagen, 2011, length 49 mm; **I, J**= *Pseudadusta (Bithloa) gumbinneri* Petuch and Drolshagen, 2011, length 39 mm.

Figure 7.21 Mollusks of the upper Fruitville Member (Unit 2), Tamiami Formation.
A= *Sinistrofulgur robesonense* (Gardner, 1948), length 225 mm; **B**= *Pyruella sarasotaensis* Petuch, 1982, length 71 mm; **C**= *Murexiella petuchi* E. Vokes, 1994, length 12 mm; **D**= *Scaphella danielleae* Petuch, 1994, length 77 mm; **E**= *Melongena sarasotaensis* Petuch, 1994, length 72 mm; **F**= *Fenestraconus myakka* Petuch and Drolshagen, 2011, length 44 mm; **G**= *Strombus jenniferae* Petuch and Drolshagen, 2011, length 70 mm; **H**= *Nodipecten vaccamavensis* (Olsson, 1914), length 74 mm; **I**= *Polygona duerri* (Petuch, 2004), length 55 mm; **J**= *Hystrivasum hyshugari* Petuch, 1994, length 72 mm.

Figure 7.22 Cowries of the upper Fruitville Member (Unit 2, Kissimmee Facies), Tamiami Formation.
A, B= *Akleistostoma (Akleistostoma) ingrami* Petuch and Drolshagen, 2011, length 53 mm; **C, D**= *Akleistostoma (Ingramicypraea) glennae* Petuch and Drolshagen, 2011, length 33 mm; **E, F**= *Akleistostoma (Olssonicypraea) zahinai* Petuch and Drolshagen, 2011, length 32 mm; **G, H**= *Pahayokea (Kissimmecypraea) lybrandi* Petuch and Drolshagen, 2011, length 83 mm; **I, J**= *Okeechobea fortdrumensis* Petuch and Drolshagen, 2011, length 42 mm.

Figure 7.23 Mollusks of the upper Fruitville Member (Unit 2, Kissimmee Facies), Tamiami Formation.

A, B= *Malea hyaducki* Petuch and Berschauer, 2021, length 124 mm; **C, D=** *Okeechobea (Yeehawcypraea) lindajoyceae* Petuch and Drolshagen, 2011, length 35 mm; **E=** *Pliculofusus seminole* (Petuch, 1994), length 130 mm; **F, G=** *Pahayokea mauriceguyi* Petuch, Berschauer, and Myers, 2018, length 70 mm; **H=** *Scaphella brennmortoni* Olsson and Petit, 1964, length 98 mm (also found in the Waccamaw Formation of North Carolina); **I=** *Hystrivasum hertweckorum* Petuch, 1994, length 85 mm; **J=** *Volutifusus kissimmeensis* Petuch & Berschauer, 2021, length 105 mm

Figure 7.24 Unit 2 Fruitville Member (Kissimmee Facies) mollusks shared with the Chowan River and Waccamaw Formations of the Carolinas.
A= *Busycon auroraensis* Petuch, 1994, length 172 mm; **B=** *Sinistrofulgur adversarium* (Conrad, 1863), length 181 mm; **C=** *Busycotypus concinnum* (Conrad, 1873), length 136 mm; **D=** *Brachysycon canaliferum* (Conrad, 1862), length 162 mm; **E=** *Mansfieldostrea ravenelliana* (Tuomey and Holmes, 1852), length 96 mm; **F=** *Volutifusus auroraensis* Petuch, 1994, length 110 mm; **G=** *Volutifusus typus* Conrad, 1863, length 138 mm; **H=** *Carolinapecten senescens* (Dall, 1898), length 87 mm.

CHAPTER 8. The Caloosahatchee Subsea (Caloosahatchee Formation)

The last million years of the Pliocene was a time of rapid sea level rises and falls and gradual climatic cooling (see Figure 4.2, Chapter 4). Along the eastern coast of the U.S. at this time, many widespread tropical Kissimmean mollusks, such as *Contraconus* and *Pterorhytis*, began to die out, probably in response to the beginning of colder water conditions (Petuch, 1994; 1995). This regional extinction is best seen by a comparison of the faunas of the latest Pliocene Chowan River Formation and the earliest Pleistocene James City Formation (Virginia and North Carolina), where these and many other tropical groups disappear at the boundary between the formations. At around 2.5 million years B.P., sea levels plummeted to the lowest they had been since the Zanclean (Figure 4.2), and the Tamiami Subsea ceased to exist. For over 500,000 years, the Okeechobean Sea basin was emergent and probably housed a series of large freshwater lakes within the Everglades depression. During the beginning of the Gelasian Age of the Pleistocene, around 2.6 million years B.P., sea levels rose abruptly, and the Okeechobean basin was reflooded with a new marine world, the **Caloosahatchee Subsea** (named for the Caloosahatchee Formation; see Petuch, 2004) (Figure 5.1). By this time, the Okeechobean Sea had become essentially landlocked, resembling a saltwater lake surrounded by low islands covered by mangroves and pine forests. Whereas the areas farther north along the Atlantic Coastal Plain had cooler water conditions, the Caloosahatchee Subsea remained warm and tropical due to solar heating of the shallow enclosed basin. Because of this localized, isolated tropical marine environment, the Caloosahatchee Subsea acted as a refugium for many of the regionally extinct Yorktown- and Duplin-type molluscan genera. Whereas groups such as *Contraconus* died out in the northern areas, they flourished within the Caloosahatchee Subsea, producing new endemic species radiations.

Deposition within the Caloosahatchee Subsea. By the beginning of Caloosahatchee time, the pseudoatoll reef tracts had become covered with sand and rubble and had become large island chains (Figure 8.1). In the west, Immokalee Island had enlarged to cover roughly one third of the width of the Hendry Platform. Likewise, the narrow Miami and Palm Beach Reef Tracts of Tamiami time had now widened and supported a long chain of islands, the Miami Archipelago. The Cape Sable Bank was also enlarging at this time and had almost fused to the southern part of Immokalee Island. A new feature, **Miccosukee Island** (named for the Miccosukee Tribe; see Petuch, 2004) had now developed along the eastern edge of the Hendry Platform. This elongated island was covered with mangrove forests and was bounded by coral reefs along its eastern side, adjacent to the deep Loxahatchee Trough. As in previous subseas, the pseudoatoll feature was separated from the Florida mainland by the narrow Caloosahatchee Strait in the west and the Loxahatchee Strait in the east. The Florida mainland adjacent to the Caloosahatchee Subsea had also become geomorphologically complex during the Plio–Pleistocene boundary time. On the east coast, the most prominent feature was the **Nashua Lagoon System** (named for the Nashua Formation; see Petuch, 2004), a series of narrow, muddy coastal lagoons that

Figure 8.1 Simulated satellite image (altitude: 100 miles) of the Floridian Peninsula during the Gelasian Pleistocene, showing the probable appearance of the **Caloosahatchee Subsea** of the Okeechobean Sea. At this time, Immokalee Island had widened and supported mangrove forests along its eastern side. The eastern edge of the Hendry Platform now contained **Miccosukee Island**, a large, elongated land mass that accreted along its eastern side. This central island supported large coral reef complexes, especially along its eastern side, adjacent to the deep Loxahatchee Trough. The muddy coastal estuaries of the **Nashua Lagoon System** can be seen to have extended all along the eastern coast of Florida, from the northern end of the Kissimmee Embayment and St. Lucie Peninsula northward to Georgia.

formed behind sandy barrier islands (extending northward to the Georgia border; see Chapter 9).

On the western coast of the Floridian Peninsula, the Myakka Lagoon System was now filling rapidly, being replaced by peaty, organic-rich mud flats, cypress forests, and freshwater marshlands. To the south, in the present-day Peace River Valley, a new broad, bay-like estuarine feature had formed. This feature, which extended to the east of present-day Arcadia, contained extensive quartz sand flats, Turtle Grass beds, and deep tidal channels filled with coral bioherms. The Okeechobean Sea basin, southern Kissimmee Embayment, and the Arcadia estuaries were the main centers of deposition for the Caloosahatchee Formation and its four members, the Fordville, Fort Denaud, Bee Branch, and Ayers Landing. The east coast Nashua Lagoon System was the center of deposition for the contemporaneous Nashua Formation and its two members, the Fort Drum and the Rucks Pit.

Caloosahatchee Formation, Okeechobee Group. The beds exposed at the low-water line along the Caloosahatchee River near Lake Okeechobee have the distinction of being the first geologic unit described from the Everglades area. First noticed and written about by Angelo Heilprin (1886), they were later given the name Caloosahatchee beds by W.H. Dall (1887 and 1892: 140–149). Later, in 1909, Matson and Clapp named them the Caloosahatchee Marl and also described and named the contemporaneous Nashua Marl. It was not until 1958 that the formation was formally described by Jules DuBar. This large, comprehensive work was DuBar's doctoral dissertation, and it contained the first detailed lithologic and paleontologic descriptions, the descriptions of new members, and a review of the depositional paleoenvironments. DuBar followed the convention of the time and referred to the unit as the Caloosahatchee Marl. We follow Petuch and Roberts (2007: 119-121) in referring to this unit as the Caloosahatchee Formation.

Lithologic Description of the Caloosahatchee Formation. Taken from DuBar (1958: 34–35): "...Typically, beds of the Caloosahatchee formation consist of marls composed primarily of quartz sand, silt, and shells. Most of the strata are soft or only slightly indurated, but some are calcareous and very hard, so as closely to approach the nature of true limestone. Most layers are moderately to abundantly fossiliferous, although some, especially sands, are almost or completely barren. Fresh exposures are generally light colored, with white, light gray, cream, and buff predominating. In the subsurface, many sand layers are light green to olive-green. Weathered marls are usually medium to dark gray." DuBar's "marls" are actually arenaceous or argillaceous calcarenites and calcilutites. These are often interbedded with quartz sand units, frequently packed with molluscan and coral bioclasts. At the Brantley Pit near Arcadia, all four members of the Caloosahatchee Formation were exposed. Lithologies within these units varied from blue-gray sandy lime mud, to light yellow-tan quartz sand with mollusk shells, to white quartz sand mixed with dolosilts. No limestones were present at the quarry, although sandy limestones were reported at quarries 20 km to the south. Other diagnostic lithologies present in the Caloosahatchee include sandy lime muds (arenaceous calcilutites) with large, dense, gray sandy concretions, and peaty sands, often with freshwater snails and clams (naiads), and carbonized wood.

The areal extent of the Caloosahatchee Formation varies with member and reflects sea level fluctuations during Plio–Pleistocene boundary time. The Fordville Member (Figure 8.2), Fort Denaud Member (Figure 8.5), and Bee Branch Member (Figure 8.15) all have, essentially, the same areal distributions, with the greatest extent being during Fort Denaud time. At its greatest development, Caloosahatchee deposition extended from southern Pinellas County to Monroe County and from southern Highlands and Okeechobee Counties to Dade County. The formation is absent under parts of the Immokalee Rise and Atlantic Coastal Ridge, being represented by undifferentiated sands. During Caloosahatchee time, these areas were islands with terrestrial conditions, which explains their lack of marine deposition. The Caloosahatchee Formation varies in thickness from 3 m at Fort Denaud, Hendry County (DuBar, 1974: Figure 4), to 10 m at the Brantley Pit near Arcadia, DeSoto County, to 15 m at the 10-Mile Bend on the Miami Canal, Palm Beach County, to over 60 m in the Loxahatchee Trough area west of Delray Beach, Palm Beach County (Puri and Vernon, 1964). The surface contours of the Caloosahatchee conform to the bathymetric contours of the Caloosahatchee Subsea. At the Palm Beach Aggregates quarries, in the Loxahatchee Trough area, the formation lies 15–18 m below surface. Along the Miami Canal, bordering the eastern edge of Miccosukee Island, the formation lies only 3 m below surface. At some areas, such as Fort Denaud on the Caloosahatchee River, the formation is less than 2 m below surface and is exposed at the lowest water line during periods of drought.

On the sea level curves shown on Figure 5.2 (Chapter 5) and on Figure 10.2 (Chapter 10), the four Caloosahatchee members can be seen to correspond to individual marine transgressions. These incursions were separated from each other by dry, regressive intervals of 100,000–200,000 years, producing obvious erosional unconformities between the individual members. The final reflooding of the Caloosahatchee Subsea during Ayers Landing time coincided with a period of low sea levels (Figure 10.2, Chapter 10). Because of this, the last Caloosahatchee deposition was much more restricted in distribution (Figure 8.18), covering only about 70% of the area of the underlying Bee Branch Member.

Stratotype. Until now, there has been no true stratotype for the Caloosahatchee Formation, only broad areas that contain exposures along the low water line of the Caloosahatchee River. Two researchers' observations: " … The typical exposures of the formation are on the Caloosahatchee River at and near LaBelle, Hendry County …" (Cooke, 1945: 215) and " … The type exposures of the Caloosahatchee marl are found along the Caloosahatchee River between Fort Denaud and Ortona Locks …" (DuBar, 1958: 35). Petuch and Roberts (2007: 120) designated the stratotype to be the section described by DuBar (1958: 236–237) at his station A36 (NW ¼ SW ¼, Sec. 11, T. 45 S., R. 28 E., Hendry County, Florida) on the " … left bank of the Caloosahatchee River about 0.9 mile upstream (east) from the bridge at Fort Denaud." The Caloosahatchee stratotype section includes DuBar's units 1–7 (with 8 being the Fort Thompson Formation and 9 and 10 being the Pamlico Formation). This section is still available for study, and crops out at 1 m below the top of the section that extends to the water line.

Age and Correlative Units. Based on the presence of key widespread index fossils such as the gastropods *Ximeniconus waccamawensis*, *Ventrilia betsiae*, *Ventrilia elizabethae*, *Sinum multiplicatus*, *Ilyanassa wilmingtonensis*, and *Ilyanassa granifera* (see Petuch, 1994) and the bivalves *Mercenaria permagna*, *Macrocallista greeni*, and *Dinocardium hazeli*, the Caloosahatchee Formation is now known to span the entire Gelasian Age of the Pleistocene. The Fordville Member is earliest Gelasian, the Fort Denaud Member is mid-Gelasian, and the Bee Branch and Ayers Landing Members are both late Gelasian. The formation is correlative with the Colerain Beach Member of the Chowan River Formation (Fordville Member only) and the James City Formation of Virginia and northern North Carolina, the Bear Bluff and Waccamaw Formations of southern North Carolina and northern South Carolina, and the Fort Drum and Rucks Pit Members of the Nashua Formation of eastern Florida.

Caloosahatchee Formation Index Fossils. As mentioned previously in this chapter, the enclosed warm Caloosahatchee Subsea acted as a refugium for many Tamiami Subsea molluscan genera. Those that survived the extinction event at the end of the Piacenzian underwent new species radiations, producing sets of highly endemic faunas. Some of the Tamiami survivors that speciated during the times of the four Caloosahatchee members include the gastropod genera *Pyrazisinus*, *Okeechobea*, *Pahayokea*, *Siphocypraea*, *Macrostrombus*, *Echinofulgur*, *Pliculofusus*, *Turbinella*, *Seminoleconus*, *Osceolaconus*, and *Contraconus*. Successive species radiations from each of these genera, together, give the fauna of each member its own distinctive appearance. The principal index species of the Caloosahatchee Formation are illustrated in iconographies at the end of the descriptions of each of the four members.

Fordville Member, Caloosahatchee Formation. The senior author, in describing the stratigraphy at the Ashland Petroleum and Asphalt Corporation (APAC) pit in Sarasota (Petuch, 1982), first noticed that the uppermost fossiliferous bed (Unit 1) contained a molluscan fauna that was different from those contained in the layers below. A more detailed analysis showed that these fossils belonged to the Caloosahatchee Formation and that the sediments ("beach deposit") were also quite different from those of the underlying Tamiami Formation (Figure 8.2). The locality of this measured section (Figure 5.3, Chapter 5) was on the eastern side of the APAC pit, and this Caloosahatchee layer was thin (1 m) and poorly developed. Subsequently, William Lyons encountered the same unit on the western side of the APAC pit, but there it was much better developed, being over 4 m thick and containing nine distinct beds (Lyons, 1992: 133–159). Although containing numerous classic Caloosahatchee index fossils such as *Turbo rhectogrammicus*, *Cerithioclava caloosaensis*, *Ximeniconus waccamawensis*, and *Cancellaria conradiana*, Unit 1 was also found to contain a large number of very distinct restricted species. Some of these were misidentified by Lyons, who gave these endemic forms the names of living Recent species (Lyons, 1992: Figure 1 to Figure 80). This newly discovered fauna has also been found elsewhere around the Everglades area, demonstrating that Unit 1 has a wider distribution than originally thought.

Based on the distinctive and complex lithology, the wide areal extent, and the chronologically restricted molluscan fauna, Petuch and Roberts (2007: 121-122) named this lowermost set of Caloosahatchee beds the Fordville Member. Their member takes its name from Fordville, a small agricultural hamlet (Mennonite farms) 3 km southeast of the APAC pit on Fruitville Road. The Quality Aggregates quarries,

Figure 8.2 View of an exposure of the Fordville Member of the Caloosahatchee Formation (Unit 1) in the APAC Pit, Sarasota, Sarasota County, Florida. This section is the **stratotype of the Fordville Member**, and the shell fragment and sandy peat layers that characterize the member can be seen to be overlain by a thick surficial unit composed of undifferentiated late Pleistocene sand.

on the other side of I-75 from the APAC pit (east), are adjacent to Fordville and also contain exposures of the new member. Because this new member is present at all the quarries in the Fordville area, and because the names Fruitville and Sarasota have already been used for other older units, the village was chosen as the only viable possibility for a formal geologic name.

Lithologic Description and Areal Distribution. As shown by Lyons (1992: 135), the lithology of the Fordville Member is quite variable. Typically, the member is composed of quartz sands with varying amounts of peats, carbonized wood, and particulate organic matter. Some quartz sand and peat layers are packed with well-preserved mollusks, often the bivalves *Carditamera*, *Anomalocardia*, and *Mulinia*.

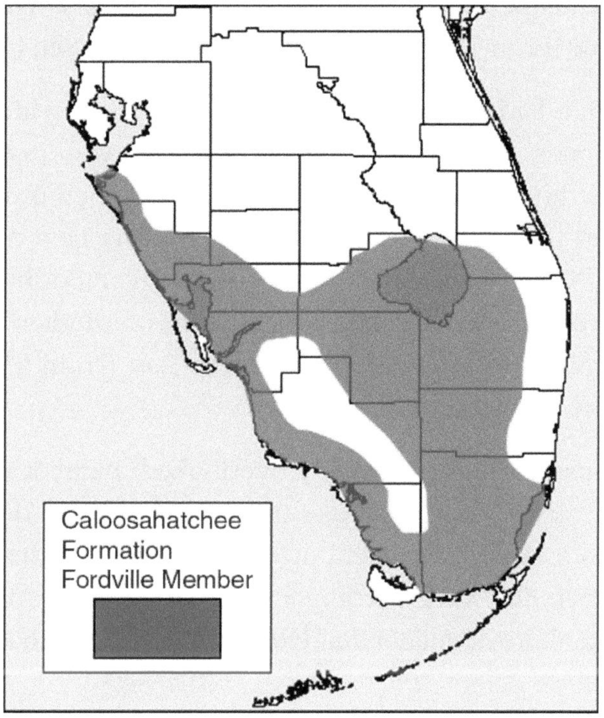

Figure 8.3 Areal extent of the Fordville Member of the Caloosahatchee Formation, in southern Florida. Throughout its distribution, the Fordville is represented by scattered, discontinuous erosional stringers.

Large calcareous biohermal masses of sabellariid colonial polychaete worms are often present, forming discontinuous reef-like structures. The quartz sands are variably colored, ranging from white, to brown, to black. Some sand units contain minor amounts of clays, usually sepiolite and illite. The mollusks and sabellariid "worm rock" masses are colored brown or blackish-brown, often weathering to a light reddish-brown. Fordville beds in the southern areas of its distribution contain a higher carbonate fraction, usually variable amounts of calcilutites and dolosilts. These beds typically contain large molluscan bioclasts and are colored a light gray or blue-green. Limestones are absent in this member. As can be seen on Figure 5.2, the Fordville Member covers a smaller area than do the subsequent Fort Denaud and Bee Branch Members. Over its range, the Fordville varies in thickness from 4 m at the stratotype at Sarasota, to 2 m in the Brantley Pit, Arcadia, DeSoto County, to 5 m along the Miami Canal (canal dredging), Palm Beach County. In the Fordville area of Sarasota, the member is only 1–3 m below surface whereas at the Brantley pit, it occurs at depths of 15 m below surface. Along the Miami Canal, the member occurs at 15–20 m below surface and at Mule Pen Quarry at East Naples, Collier County, it lies 5–8 m below surface.

Stratotype. Petuch and Roberts (2007: 122) designated Unit 1 on the western side of the APAC pit (see Figure 5.3, Chapter 5) as the stratotype of the Fordville Member. This site, located on Richardson Road, adjacent to I-75, is now flooded as a lake in a Sarasota County public park. A map of the stratotype (T. 36 S., R. 18 E., Sec. 12, Sarasota County) was given by Lyons (1992: Figure 1). This locality is approximately 3 km northwest of Fordville, Sarasota County (on Fruitville Road). The Fordville stratotype beds are 2–3 m below the mean lake surface and can be accessed by scuba diving.

Age and Correlative Units. Based on the presence of key widespread index fossils, such as the gastropods *Pterorhytis conradi*, *Littorinopsis lindae*, and *Pliculofusus cronleyensis* and the bivalve *Conradostrea sculpturata*, the Fordville Member is now known to date from the latest Piacenzian Pliocene. The new member is correlative with the Colerain Beach Member of the Chowan River Formation of Virginia and northern North Carolina, the Bear Bluff Formation of southern North Carolina and northern South Carolina, and the basal beds of the Fort Drum Member of the Nashua Formation of eastern Florida.

Fordville Index Fossils. This previously overlooked member contains a distinctive suite of chronologically restricted species that can be used to demarcate the member boundaries. Besides a large component of endemic species, the Fordville fauna also contains a small but noteworthy component of upper Chowan River Formation species. Some of the more abundant and indicative Fordville index taxa are shown on the following figure.

Fort Denaud Member, Calooshatchee Formation. In his 1958 study of the Caloosahatchee, DuBar recognized three distinct sections: an "upper shell bed," a middle Bee Branch Member (limestone), and a "lower shell bed" (DuBar, 1958: 50–64, Figure 23). He later (1958a; 1974) designated the upper bed the Ayers Landing Member and the lower bed the Fort Denaud Member. Of the three members, the Fort Denaud is the most stratigraphically-complex and contains the richest molluscan fauna of the entire formation. This member contains the Caloosahatchee mollusk species described and illustrated by Heilprin (1886) and Dall (1890). DuBar (1958; 1974) was the first person to describe the paleoecology and depositional environments of the Caloosahatchee Formation. He noted that, of the three members, the Fort Denaud contained the most types of environments, including coral patch reefs (DuBar, 1974: 217), oyster bioherms (the "basal oyster biostrome"), mud flats containing *Cyrtopleura* and *Panope* deep-burrowing bivalves (the "*Cyrtopleura costata*

Figure 8.4 Mollusks of the Fordville Member, Caloosahatchee Formation.
A, B= *Siphocypraea streami* Petuch and Drolshagen, 2011, length 66 mm; **C=** *Pterorhytis conradi* (Dall, 1890), length 59 mm; **D=** *Pterorhytis seminola* Olsson and Petit, 1964, length 27 mm; **E=** *Hexaplex trippae* Petuch, 1994, length 81 mm; **F=** *Hystrivasum martinshugari* Petuch, 1994, length 92 mm; **G=** *Americoliva briani* Petuch, 1994, length 86 mm; **H=** *Pliculofusus cronleyensis* (Gardner, 1948), length 152 mm (complete specimen from Aurora, North Carolina; only fragmentary specimens have been collected in Florida); **I=** *Pusula permagna* Johnson, 1910, length 22 mm.

Faunizone"), *Vermicularia* worm gastropod bioherms, brackish water calcarenite beds with *Rangia* clams, and freshwater environments.

Lithologic Description and Areal Distribution. From DuBar (1958: 51), for his "lower beds," later named the Fort Denaud Member: "Typically, the lower beds are light colored, with cream, white, and light gray predominating, although some units are mottled yellow-brown by limonite staining, and all are some shade of gray on weathered surfaces. At almost all localities, the beds are comprised of sandy (quartz) and silty marls. Locally, the percentages of sand-sized particles is high enough to warrant describing the rock as calcareous sandstone ... Calcareous concretions of various sizes and shapes are abundant. Many are small fossiliferous nodules that occur in random manner scattered in the bed, whereas some are joined so as to form an arborescent network extending throughout an entire bed ... Many of the beds are fossiliferous. Almost every facies and all units carry an abundant, varied, and well preserved fauna. Mollusks are dominant, but foraminifers are well represented, and ostracodes, barnacles, and bryozoans are common." As discussed earlier in the chapter, the Fort Denaud has the widest areal extent of all the Caloosahatchee members (Figure 8.5), extending from Pinellas, southern Highlands, southern Okeechobee, and southern Martin Counties in the north to Monroe and Dade Counties in the south. Throughout its range, the Fort Denaud is also the thickest member, being over 10 m thick at St. Petersburg, Pinellas County, 4 m thick along the Miami Canal in Palm Beach County, 30 m thick west of Delray Beach, Palm Beach County, and 1–3 m thick along the central area of the Caloosahatchee River.

Stratotype. From DuBar (1958: 50–51): "... Exposures of the lower beds (Fort Denaud Member) are most typically developed along a stretch of the river (Caloosahatchee) extending between points 0.5 and 3.5 mi upstream from Fort Denaud in Hendry County. Here, the beds rise higher above the water level than elsewhere, are usually thicker, and are laterally more extensive." Petuch and Roberts (2007: 125) selected one of DuBar's Fort Denaud area collecting localities (A39) as the stratotype (SW ¼, NW ¼, Sec. 10, T. 43 S., R. 28 E., Hendry County, Florida): " ... right bank of the Caloosahatchee River about 325 yards downstream from the bridge at Fort Denaud and 170 yd downstream from Jack's Branch." Here, the Fort Denaud occurs at 3 m below the top of the section.

Age and Correlative Units. Based on the presence of widespread key index fossils, such as the gastropods *Ximeniconus waccamawensis*, *Ilyanassa granifera*, *Ilyanassa wilmingtonensis*, *Globinassa schizopyga*, *Sinum multilineatum*, and *Triplofusus acmaensis*, and the bivalves *Argopecten vicenarius* and *Carolinapecten solaroides*, the Fort Denaud Member is now known to date from the mid- Gelasian Age, at approximately 2.2 million years B.P. (shown in Petuch, 1994; 2004). The member is correlative with the lower beds of the James City Formation of Virginia and northern North Carolina, the lower beds of the Waccamaw Formation of southern North

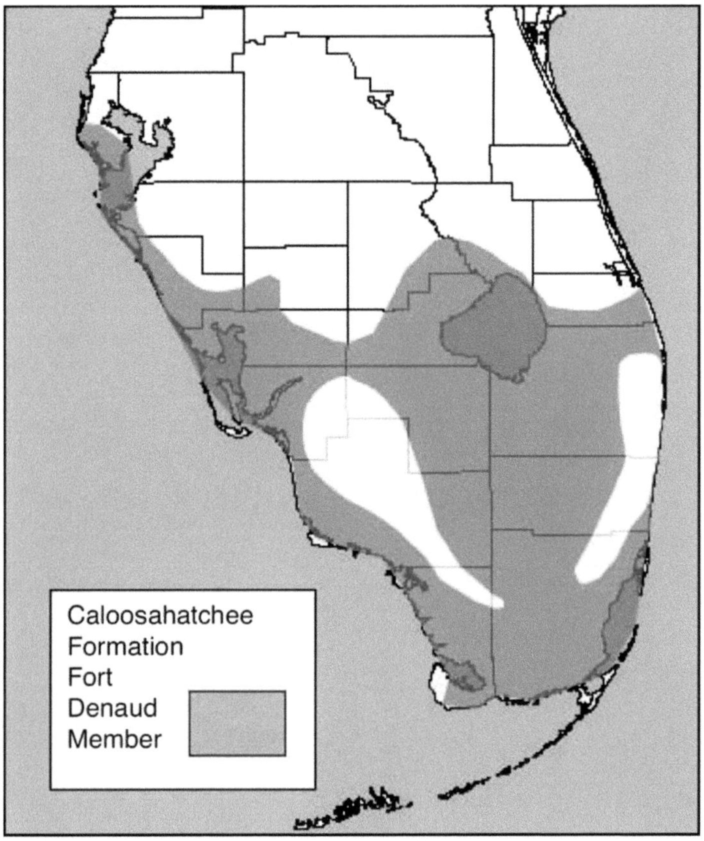

Figure 8.5 Areal extent of the Fort Denaud Member, Caloosahatchee Formation, in southern Florida.

Carolina and northern South Carolina, and the upper beds of the Fort Drum Member of the Nashua Formation of eastern Florida.

Fort Denaud Index Fossils. DuBar (1974: Table 5) gave a listing of characteristic macrofossils for his Caloosahatchee members, showing that many species were present in all three. Subsequent studies by the senior author (Petuch, 1991; 1994; Petuch and Roberts, 2007) have shown that many of DuBar's chronologically widespread taxa actually represented successions of chronologically restricted descendant species that evolved rapidly from their Fort Denaud ancestors. Although containing a small component of widespread Waccamaw and James City species, the majority of the Fort Denaud molluscan and coral faunas are composed of Floridian endemics, restricted to the Caloosahatchee Subsea. Some of the more prominent and characteristic index species are shown here in the following iconography.

Bee Branch Member, Caloosahatchee Formation. During DuBar's stratigraphic survey of the Caloosahatchee River, he recognized a distinct, prominent limestone layer midway in the section. This resistant bed frequently formed a wide, thick ledge that jutted out from the section just above the low-water line. The best-developed exposures were found to be in the vicinity of the confluence of a small tributary of the Caloosahatchee River, the Bee Branch, and the member name derives from this stream. Petuch (1994) recognized Bee Branch sediments and fossils from the Miami

Figure 8.6 Cowries of the Fort Denaud Member, Caloosahatchee Formation.
A, B= *Siphocypraea problematica* Heilprin, 1886, length 62 mm; **C, D=** *Siphocypraea (Floridacypraea) dubari* Petuch and Drolshagen, 2011, length 78 mm; **E, F=** *Siphocypraea (Lokossea) swearingeni* Petuch and Drolshagen, 2011, length 86 mm; **G, H=** *Pahayokea josiai* (Fehse, 1997), length 42 mm; **I, J=** *Okeechobea philemoni* (Fehse, 1997), length 61 mm.

Figure 8.7 Gastropods of the Fort Denaud Member, Caloosahatchee Formation.
A, B= *Okeechobea shawnae* Petuch and Drolshagen, 2011, length 68 mm; **C, D=** *Cassis calusa* Petuch and Berschauer, 2018, length 154 mm; **E=** *Semicassis alligator* (Petuch, 1991), length 47 mm; **F=** *Macrostrombus leidyi* (Heilprin, 1886), length 147 mm; **G=** *Cerithioclava caloosaense* (Dall, 1892), length 68 mm; **H=** *Strombus keatonorum* Petuch, 1994, length 73 mm; **I=** *Pliculofusus scalarinus* (Heilprin, 1886), length 160 mm.

Figure 8.8 Busyconid Gastropods of the Fort Denaud Member, Caloosahatchee Formation. **A**= *Pyruella eismonti* Petuch, 1994, length 115 mm; **B**= *Laevisycon planulatum* (Dall, 1890), length 65 mm; **C**= *Laevisycon soror* (Petuch, 1994), length 72 mm; **D**= *Busycon duerri* Petuch, 1994, length 217 mm; **E**= *Sinistrofulgur palmbeachensis* Petuch, 1994, length 282 mm; **F**= *Sinistrofulgur caloosahatcheensis* Petuch, 1994, length 92 mm; **G**= *Busycoarctum rapum* (Heilprin, 1886), length 177 mm; **H**= *Fulguropsis floridanum* (Olsson and Harbison, 1953), length 112 mm.

Figure 8.9 Gastropods of the Fort Denaud Member, Caloosahatchee Formation.
A= *Phyllonotus evergladesensis* Petuch, 1994, length 86 mm; **B**= *Pterorhytis wilsoni* Petuch, 1994, length 40 mm; **C**= *Vokesinotus lepidotus* (Dall, 1890), length 22 mm; **D**= *Vokesimurex pahayokee* Petuch, 1994, length 34 mm; **E**= *Eupleura calusa* Petuch, 1994, length 33 mm; **F**= *Heilprinia caloosaensis* (Heilprin, 1886), length 68 mm; **G**= *Heilprinia evergladesensis* Petuch, 1994, length 104 mm; **H**= *Fasciolaria monocingulata* Dall, 1890, length 78 mm; **I**= *Murexiella macgintyi* (M. Smith, 1938), length 25 mm; **J**= *Knefastia lindae* Petuch, 1994, length 57 mm.

Figure 8.10 Gastropods of the Fort Denaud Member, Caloosahatchee Formation.
A= *Melongena caloosahatcheensis* Petuch, 1994, length 122 mm; **B=** *Cerithidea xenos* Petuch, 1994, length 25 mm; **C=** *Cerithidea jenniferae* Petuch, 1994; length 27 mm; **D=** *Pyrazisinus scalatus* (Heilprin, 1886), length 78 mm; **E=** *Pyrazisinus ecarinatus* (Dall, 1892), length 61 mm; **F=** *Nerita hertweckorum* Petuch, 1994, length 17 mm; **G=** *Scaphella floridana* Heilprin, 1886, length 145 mm; **H=** *Turbinella regina* Heilprin, 1886, length 286 mm; **I=** *Chicoreus calusa* Petuch, 1994, length 43 mm.

Figure 8.11 Gastropods of the Fort Denaud Member, Caloosahatchee Formation.
A= *Turbinella scolymoides* Dall, 1890, length 190 mm; **B**= *Contraconus tryoni* (Heilprin, 1886), 158 mm; **C**= *Osceolaconus osceolai* (Petuch, 1994), length 88 mm; **D**= *Calusaconus spuroides* (Olsson and Harbison, 1953, length 37 mm; **E**= *Gradiconus parkeri* (Richards and Harbison, 1947), length 26 mm; **F**= *Porphyria paraporphyria* (Petuch, 1991), length 67 mm; **G**= *Americoliva roseae* (Petuch, 1991), length 83 mm; **H**= *Americoliva erici* (Petuch, 1994), length 87 mm; **I**= *Acantholabia floridana* Olsson and Harbison, 1953, length 30 mm.

Figure 8.12 Mollusks of the Fort Denaud Member, Caloosahatchee Formation.
A= *Carolinapecten solaroides* (Heilprin, 1886), length 172 mm; **B=** *Stralopecten caloosaensis* (Dall, 1898), length 65 mm; **C=** *Hystrivasum horridum* (Heilprin, 1886), length 99 mm; **D=** *Liochlamys bulbosa* (Heilprin, 1886), length 95 mm; **E=** *Echinofulgur echinatum* (Dall, 1890), length 88 mm; **F=** *Pleioptygma lineolata* (Heilprin, 1886), length 116 mm; **G=** *Caloosarca crassicosta* (Heilprin, 1886), length 58 mm; **H=** *Triplofusus acmaensis* (B. Smith, 1940) (with encrusting *Septastrea* coral), length 358 mm; **I=** *Chicoreus sarae* Petuch, 1994, length 57 mm.

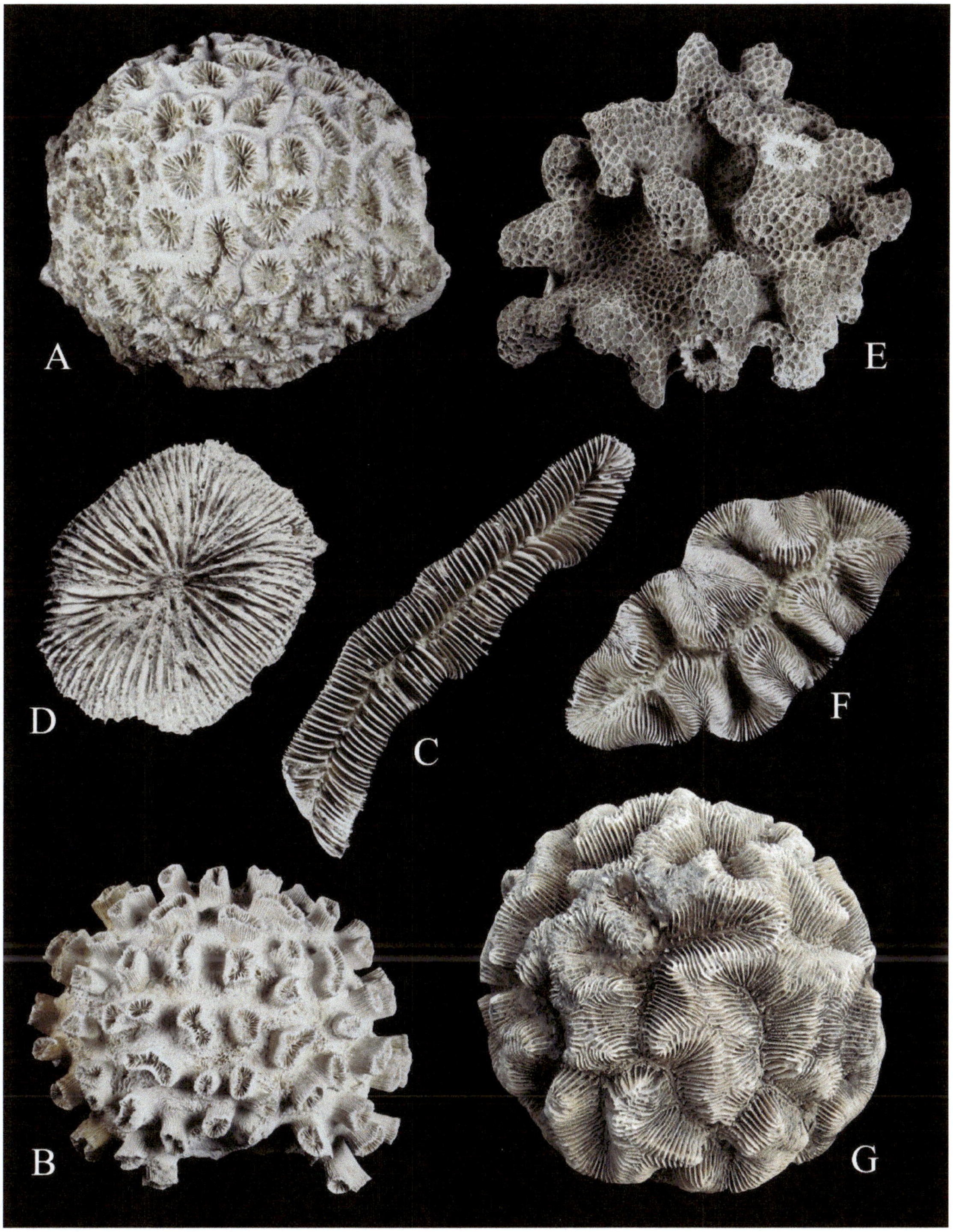

Figure 8.13 Corals of the Fort Denaud Member, Caloosahatchee Formation (Coral Bioherms of the Caloosahatchee Subsea).
A= *Dichocoenia caloosahatcheensis* Weisbord, 1974, length 76 mm; **B**= *Dichocoenia eminans* Weisbord, 1974, length 146 mm; **C**= *Thysanus floridanus* Weisbord, 1974, length 141 mm; **D**= *Scolymia* cf. *lacera* (Pallas, 1766), length 75 mm; **E**= *Oculina* species, length 94 mm; **F**= *Manicina pliocenica* Gane, 1900, length 146 mm; **G**= *Manicina puntagordensis* Weisbord, 1968, length 127 mm. These corals were abundant on the reefs off Miccosukee Island.

Figure 8.14 Reef-Associated Gastropods of the Fort Denaud Member, Caloosahatchee Formation. **A, B=** *Malea springi* Petuch, 1989 length 186 mm; **C=** *Purpuriconus protocardinalis* (Petuch, 1991), length 27 mm; **D=** *Subpterynotus gabbi* Petuch and Berschauer, 2020, length 66 mm; **E=** *Jenneria richardsi* Olsson, 1967, length 28 mm; **F=** *Cyphoma pacei* Petuch and Drolshagen, 2011, length 25 mm; **G, H=** *Oniscidia macgintyi* (M. Smith, 1937), length 20 mm; **I=** *Babelomurex mansfieldi* (McGinty, 1940), length 20 mm (the true *B. mansfieldi* is only known as a Caloosahatchee fossil; similar living species referred to this name are not this shell but represent other taxa such as *B. scalariforme*, *B. cariniferus*, etc.). All of these gastropods were common on the reefs off Miccosukee Island.

Canal dredgings, Palm Beach County, and later found large exposures in the Brantley Pit near Arcadia, DeSoto County.

Lithologic Description and Areal Distribution. From DuBar (1958: 58, condensed here): "The name Bee Branch member is here proposed to designate a relatively hard, solution-riddled marine limestone or marl unit of the Caloosahatchee marl. It is typically exposed along the Caloosahatchee River in Hendry County near the confluence with the tributary called Bee Branch … Two principal facies of the Bee Branch member and many minor variations of each have been recognized. One facies is represented by the Bee Branch deposits of the type locality which form a massive, hard calcareous bed … It is riddled with large and small solution holes that are filled with matrix from overlying beds. Commonly, small pockets of softer marly material occur within the bed … Most of the weathered surfaces are dark gray or black, but fresh surfaces are buff or cream colored. Yellow-brown mottling by limonite stain is common … The other dominant facies differs from the type locality in being softer, less consolidated, more arenaceous (quartz sand), and more distinctly concretionary. Generally, it does not form prominent ledges. The concretions contained in it are hard, dense calcareous nodules which locally occur in isolated manner but more commonly form an arborescent network extending throughout the entire bed."

As discussed earlier, the Bee Branch has the second-largest areal distribution of the four Caloosahatchee members (Figure 8.15). It ranges from southernmost Sarasota, western DeSoto, southeastern Highlands, southern Okeechobee, and southern Martin Counties in the north, to Monroe and Dade Counties in the south. Throughout its range, the Bee Branch Member is relatively thin, being 2–3 m thick in the Brantley Pit, near Arcadia, DeSoto County, 0.5–2 m thick along the Caloosahatchee River near LaBelle, Hendry County, to 3 m thick along the Miami Canal in western Palm Beach County.

Stratotype. DuBar (1958: 59) gave the type locality as the bank of the Caloosahatchee River in a stretch between his stations A22 and A26 in Hendry County (… left bank of the Caloosahatchee River about 0.8 mile downstream from the bridge at LaBelle" and 1025 yd downstream from the "sharp bend in the canal" (near the confluence with Bee Branch). DuBar (1974: 217) later stated that his station A70 is the stratotype of the Bee Branch but gives no details or references. Petuch and Roberts, 2007: 132) designated, as the lectostratotype of the Bee Branch Member, DuBar's (1958) station A22 (NW ¼ Sec. 5, T. 43 S., R. 29 E., Hendry County, Florida) at the "left bank Caloosahatchee River about 0.8 mile downstream from the bridge at LaBelle." The member is exposed at 2 m below the top of the section.

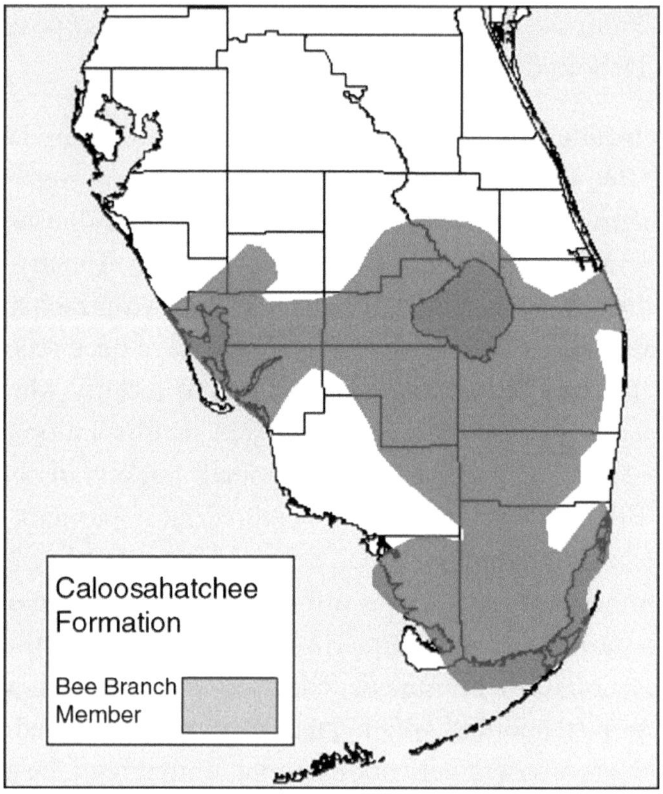

Figure 8.15 Areal extent of the Bee Branch Member, Caloosahatchee Formation, in southern Florida.

Age and Correlative Units. Based on its stratigraphic position within the Caloosahatchee Formation, the Bee Branch is here dated as being of late Gelasian Pleistocene age. The member is correlative with the upper beds of the James City Formation of Virginia and northern North Carolina, the upper beds of the Waccamaw Formation of southern North Carolina and northern South Carolina, and the lower beds of the Rucks Pit Member of the Nashua Formation of eastern Florida.

Bee Branch Index Fossils. Like the Fort Denaud Member, the Bee Branch contains a highly endemic fauna that is restricted to the Everglades area. Because much of the Bee Branch is composed of limestone, the following list is of index fossils that are primarily associated with the unconsolidated calcarenite facies. The richest Bee Branch faunas encountered to date were found along the Miami Canal, Palm Beach County, in excavations west of the Okeelanta sugar mill, and at the Brantley Pit near Arcadia, DeSoto County. Some of the more abundant and indicative Bee Branch species are shown here on the following figures.

Ayers Landing Member, Caloosahatchee Formation. In his stratigraphic survey of the Caloosahatchee River area, DuBar (1958) recognized an “upper shell bed” within the Caloosahatchee sequence. He later (1958a; 1974) named this the Ayers Landing Member, after an old boat landing near the Fort Denaud bridge. Along the river, this uppermost set of beds can be seen to lie directly on the limestone ledges of the Bee Branch Member. In the late 1980s, the senior author recognized Ayers Landing

Figure 8.16 Cowries of the Bee Branch Member, Caloosahatchee Formation.
A, B= *Siphocypraea haleyorum* Petuch and Drolshagen, 2011, length 78 mm; **C, D**= *Siphocypraea (Floridacypraea) desotoensis* Petuch and Drolshagen, 2011, length 58 mm; **E, F**= *Siphocypraea (Lokossea) wigginsi* Petuch and Drolshagen, 2011, length 83 mm; **G, H**= *Okeechobea brantleyi* (Petuch, 2004), length 55 mm; **I, J**= *Pahayokea aspenae* (Petuch, 2004), length 61 mm.

Figure 8.17 Gastropods of the Bee Branch Member, Caloosahatchee Formation.
A= *Sinistrofulgur labelleensis* Petuch, 1994, length 243 mm; **B**= *Fulguropsis elongatus* (Gill, 1867), length 118 mm; **C**= *Phyllonotus labelleensis* Petuch, 1994, length 86 mm; **D**= *Echinofulgur palmbeachensis* Petuch, 1994, length 77 mm; **E**= *Macrostrombus brachior* (Petuch, 1994), length 138 mm; **F**= *Pliculofusus labelleensis* (Petuch, 1994), length 101 mm; **G**= *Cancellaria clewistonensis* Olsson and Harbison, 1953, length 51 mm; **H**= *Fasciolaria seminole* Petuch, 1994, 64 mm; **I**= *Strombus brantleyi* Petuch and Drolshagen, 2011, length 88 mm; **J**= *Dauciconus gravesae* (Petuch, 1994), length 29 mm (*D. burnetti* is a synonym).

sediments and index fossils in excavations at the Griffin Brothers Pit adjacent to the Holey Land Wildlife Management Area in southwesternmost Palm Beach County. Many of these were later described (Petuch, 1991; 1994), and these have subsequently been collected at numerous localities, such as along the Miami Canal and at the Brantley Pit.

Lithologic Description and Areal Distribution. From DuBar (1958: 62–63, condensed here): "At many localities the bed is very arenaceous and the calcium carbonate content is mainly in the form of mollusk shells. Thick beds are generally only slightly consolidated and are easily eroded. The sands (quartz) are usually tan to yellow-brown, but the fossils are nearly white. At some localities, the unit is better consolidated, much more calcareous, concretionary, and finer grained. The color is usually light buff, but weathers to dark gray … . At all localities, the member is fossiliferous. The faunal remains consist mostly of mollusk shells, but corals, foraminifers, ostracodes, barnacles, and bryozoans are common … thin beds occur near the top of the section and appear to be of fresh water origin; other beds clearly represent a brackish water environment."

At the stratotype, the member is composed of three distinct beds: an upper tan-colored, slightly concretionary, unconsolidated marine calcarenite with sparse marine mollusks, a middle sandy, tan, freshwater limestone with freshwater snails, and a lower sandy, cream-colored marine calcarenite (weathering to gray) filled with fossil mollusks. As can be seen on Figure 5.10, the Ayers Landing Member has the smallest areal distribution of the four Caloosahatchee members. It extends from southern Sarasota, southwestern DeSoto, Charlotte, Glades, and western Palm Beach Counties in the north, to Dade County in the south. From the sea level curves shown on Figure 6.1 (Chapter 6), the Ayers Landing can be seen to have been deposited during the smallest marine transgression of the Pleistocene, and this explains its limited distribution. Over this small range, the member varies in thickness from 1 m at the Brantley Pit near Arcadia, DeSoto County, to greater than 3 m along the Caloosahatchee River in Hendry County, to 2 m at the Griffin Brothers Pit in southwesternmost Palm Beach County, to 1 m in canal digs west of the Miami Canal, west of the Okeelanta sugar mill, Palm Beach County.

Stratotype. From DuBar (1958: 62, condensed here): " … The best exposures of the shell bed occur along the Caloosahatchee River in Hendry County between stations A26 and A35, and this is considered to be the type locality." The stratotype is here recognized as DuBar's station A35, in the Ayers Landing area, "90 yards upstream from Turtle Branch" (SW ¼, NW ¼, Sec. 11, T. 43 S., R. 28 E., Hendry County, Florida). DuBar commented that at this locality " … there occurs an excellent exposure of the upper Caloosahatchee shell bed that contains, in addition to a profusion of mollusks and coral heads, a good vertebrate fauna including the teeth of *Equus (Equus)* sp. cf. *E. (E.) leidyi*."

Age and Correlative Units. Based on the stratigraphic position within the Caloosahatchee sequence, the Ayers Landing Member dates from the late Gelasian Pleistocene. The member is correlative with the uppermost beds of the James City Formation of Virginia and northern North Carolina, the uppermost beds of the Waccamaw Formation of southern North Carolina and northern South Carolina, and the uppermost beds of the Rucks Pit Member of the Nashua Formation of eastern Florida.

Ayers Landing Index Fossils. This youngest Caloosahatchee member (approximately 1.8 million years) contains the last assemblages of the Tamiami-type molluscan faunas. Some of the classic genera that appear for the last time within the Ayers Landing Member include *Contraconus*, *Siphocypraea*, *Okeechobea*, *Pahayokea*, *Liochlamys*, *Hystrivasum*, and *Echinofulgur*. Their disappearance at the end of Ayers Landing time can be used to demarcate the boundary between the Caloosahatchee Formation and the younger Bermont Formation (see Chapter 10). Like the Bee Branch Member, the Ayers Landing contains a highly endemic fauna that is restricted to the Everglades area. Some of the more abundant and indicative species are shown in the following iconography.

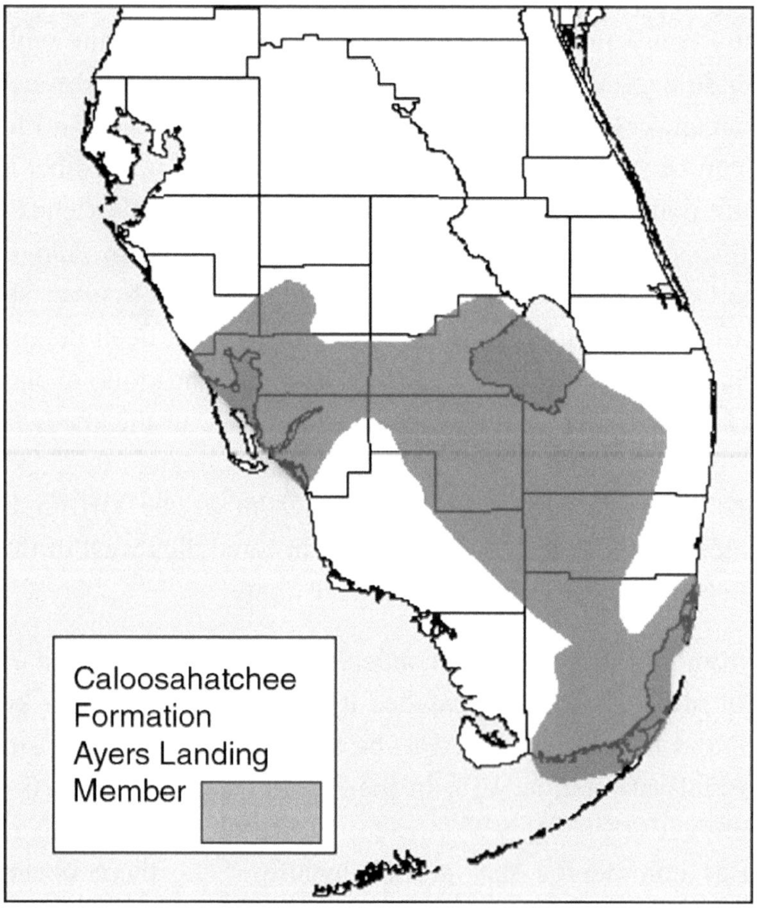

Figure 8.18 Areal distribution of the Ayers Landing Member, Caloosahatchee Formation, in southern Florida.

Figure 8.19 Cowries of the Ayers Landing Member, Caloosahatchee Formation.
A, B= *Siphocypraea ridgdilli* Petuch and Drolshagen, 2011, length 41 mm; **C, D**= *Siphocypraea (Floridacypraea) griffini* Petuch, 1991, length 39 mm; **E, F**= *Siphocypraea (Lokossea) registeri* Petuch and Drolshagen, 2011, length 65 mm; **G, H**= *Okeechobea loxahatcheensis* Petuch and Drolshagen, 2011, length 52 mm; **I, J**= *Pahayokea briskinae* Petuch, Berschauer, and Myers, 2018, length 34 mm.

Figure 8.20 Gastropods of the Ayers Landing Member, Caloosahatchee Formation.
A, B= *Cassis powelli* Petuch and Berschauer, 2018, length 195 mm; **C=** *Jenneria keatonae* Petuch, 1994, length 21 mm; **D, E=** *Oniscidia obrienae* (Olsson and Petit, 1964), length 31 mm; **F=** *Macrostrombus jonesorum* (Petuch, 1994), length 194 mm; **G=** *Turbinella lindae* Petuch, 1994, length 178 mm; **H=** *Strombus ayersensis* Petuch and Drolshagen, 2011, length 87 mm.

Figure 8.21 Gastropods of the Ayers Landing Member, Caloosahatchee Formation.
A= *Fulguropsis griffini* Petuch, 1994, length 93 mm; **B**= *Phyllonotus martinshugari* Petuch, 1994, length 58 mm; **C**= *Echinofulgur griffini* Petuch, 1994, length 105 mm; **D**= *Hystrivasum griffini* Petuch, 1994, length 66 mm; **E**= *Pyrazisinus intermedius* Petuch, 1994, length 52 mm; **F**= *Americoliva jenniferae* Petuch, 1994, length 48 mm; **G**= *Scaphella oleiniki* Petuch, 1994, length 63 mm; **H**= *Melongena (Rexmela) acutangulata* Petuch, 1994, length 47 mm; **I**= *Liochlamys griffini* Petuch, 1994, length 60 mm.

Figure 8.22 Gastropods of the Ayers Landing Member, Caloosahatchee Formation.
A= *Scaphella griffini* Petuch, 1994, length 90 mm; **B**= *Scaphella tomscotti* Petuch, 1994, length 115 mm; **C**= *Contraconus mitchellorum* Petuch, 1994, length 142 mm; **D**= *Contraconus scotti* Petuch, 1994, length 59 mm; **E**= *Contraconus heilprini* Petuch, 1994, length 32 mm; **F**= *Calusaconus charneyi* Petuch and Drolshagen, 2011, length 41 mm; **G**= *Lindaconus micanopy* Petuch, 1994, length 90 mm; **H**= *Ximeniconus robertsi* (Olsson and Harbison, 1953), length 38 mm; **I**= *Pyruella ovoidea* Petuch, 1994, length 114 mm.

Chapter 9. The Nashua Lagoon System (Nashua Formation)

The pioneer Floridian geologists Matson and Clapp first reported on a fossiliferous "white shell marl" (calcarenite) from the vicinity of Nashua, Putnam County, on the eastern side of the St. John's River south of Palatka (Matson and Clapp, 1909). Further analyses showed that this site contained some Caloosahatchee-aged fossils, and they assigned a Pliocene age to the beds within the Nashua area. Because these beds contained more clay and fewer carbonate fines than did the Caloosahatchee "marl," and because the molluscan fossils of the St. John's River beds were of a cooler water, Waccamaw-type, Matson and Clapp formally named these beds the Nashua Marl (later Nashua Formation). Some later workers, such as Cooke and Mossom (1929: 152), considered these differences insignificant and included the Nashua in the Caloosahatchee Formation. In light of new information from better sections, we here consider the Nashua to be a separate formation, contemporaneous with the Caloosahatchee Formation from farther south within the Everglades Basin.

In 1940, Sydney Stubbs of the Florida Geological Survey reported on a Nashua assemblage of fossils (as "Caloosahatchee") from a 30 m-deep well at Sanford, Seminole County. Stubbs also described similar fossils and sediment types from wells in DeLeon Springs and Daytona Beach, Volusia County, and these were cited by Cooke (1945: 227, from unpublished memoranda). More recently, the senior author (Petuch, 1994) reported on a Nashua fauna from Rucks Pit in Fort Drum, northern Okeechobee County and described several new key index fossils. A molluscan fauna and an accompanying set of lithofacies, similar to those at Fort Drum, were also found at the Dickerson Aggregates Corporation, Inc., Indrio quarry, west of Fort Pierce, St. Lucie County (Petuch, Myers, and Berschauer, 2015:64). All of these localities demonstrate that the Nashua Formation has a much larger areal distribution than was originally thought.

Lithology of the Nashua Formation. The Nashua Formation was deposited within a narrow coastal lagoon, referred to as the **Nashua Lagoon System** (Petuch, 2004:211), that had formed behind a chain of sand barrier islands. This narrow coastal waterway extended from southern Georgia all the way to St. Lucie County, Florida (see Figure 8.1 in the previous chapter). The southern end of the Nashua Lagoon, in the vicinity of present-day Indian River and St. Lucie Counties, was connected to the northern end of the Kissimmee Embayment of the Caloosahatchee Subsea by a series of narrow channels. These channels effectively cut the St. Lucie Peninsula into a chain of wide islands. The northernmost of these channels emptied into a large, shallow bay at the northern end of the Kissimmee Embayment. This was the center of deposition for the Everglades component of Nashua sedimentation. As pointed out by Cooke (1945: 214), the molluscan fossils of the Nashua Formation are cooler water species, and many of them are found in the contemporaneous Waccamaw Formation of the Carolinas. This indicates that, during Caloosahatchee Subsea time, the open Atlantic was much cooler than the interior of the enclosed Okeechobean Sea. The presence of cooler water species, mixed with some tropical forms, demonstrates that the

Figure 9.1 View of the north side of the Rucks Pit, Fort Drum, Okeechobee County, Florida, showing the three main beds of the Fort Drum Member of the Nashua Formation. These beds constitute the stratotype of the Fort Drum Member of the Nashua Formation. The upper member of the Nashua Formation, the Rucks Pit Member, had been removed from this section and sold as road fill. Photo by Anton Oleinik.

depositional environments of the Nashua were oceanographically intermediate between those of the high tropical Caloosahatchee and temperate Waccamaw and James City Formations. These cool water conditions are reflected in the complete absence of hermatypic coral reefs and bioherms and by the preponderance of terrigenous siliciclastics. The connecting channels into the Nashua Lagoon System allowed Waccamaw-type cooler water conditions, sediments, and ecosystems to exist at the northern end of the Kissimmee Embayment.

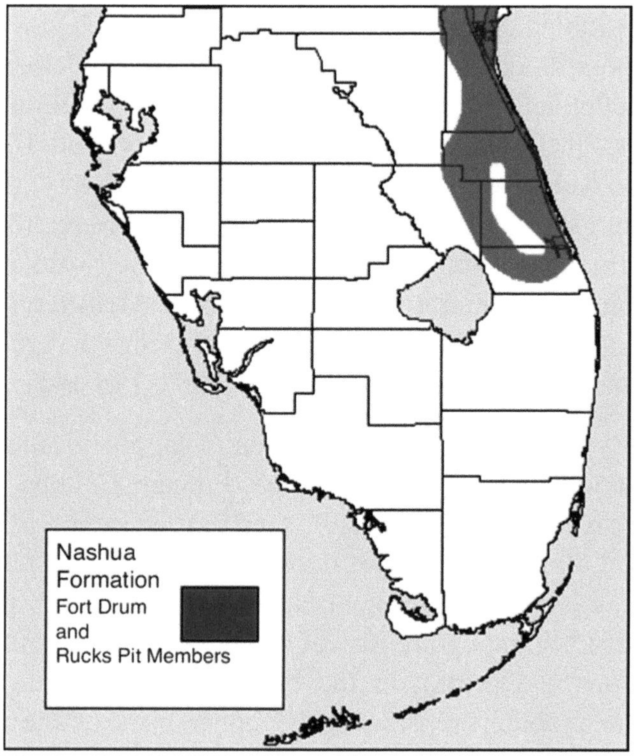

Figure 9.2 Areal distribution of the Nashua Formation and its two members, the Fort Drum and the Rucks Pit, in southern Florida.

Typically, the Nashua Formation comprises sets of interbedded units with varying lithologies, including unconsolidated sandy (quartz) calcarenites and calcilutites with variable amounts of clay minerals (mostly sepiolite and illite), quartz sand with densely packed calcareous bioclasts made up mostly of lagoonal and estuarine bivalves (Figure 9.1), shell hash coquinas with spar cementing, and cavity-filled, very sandy, spar-cemented limestone (a vuggy arenaceous sparite), at some localities grading into a sandstone. At the Rucks Pit (later known as the "Fort Drum Crystal Mine"), this sandy sparite contains abundant molluscan shell calcite geodes filled with dogtooth spar. The unconsolidated sandy calcarenites (like those at the stratotype) are white or pale gray with white or pale tan molluscan bioclasts. The spar-cemented sandy limestones (or limy sandstones) are typically dark gray with honey-gold-colored calcite crystals forming as vug filling or as molluscan geodes. True coquinas are the least common lithology, and are found in thin (less than 0.5 m thick) beds or lenses and are of a pale tan or cream color. The clay component of the unconsolidated beds increases in abundance in the northern extent of the formation. In the southern areas of Nashua deposition, the dark gray sandy limestones often occur in thick, massive units and are mined commercially for road construction.

The Nashua Formation occurs in a narrow strip that parallels the present coastline, extending inland 30–60 km (Figure 9.2). In the south, the formation extends into the northern end of the Okeechobee Plain. The Nashua ranges from St. Johns, Putnam, Flagler, Volusia, Seminole, easternmost Orange, easternmost Osceola and Brevard

Counties south to Indian River, northernmost Okeechobee, St. Lucie, and Martin Counties. Throughout its range, the Nashua is variable in thickness, being 5 m thick at the stratotype in Putnam County, 3.5 m thick at Sanford, Seminole County, 8–10 m thick at the Rucks Pit, Fort Drum, Okeechobee County, and 10–12 m thick in the Indrio Pit on Indrio Road west of Fort Pierce, St. Lucie County, the Nashua comes to within 1.5–5 m from land surface. Farther inland (to the west), the Nashua can occur at depths ranging from 20–30 m (such as at Sanford). The Nashua Formation contains two lithostratigraphic members: a lower **Fort Drum Member** (early Gelasian Age Pleistocene) and an upper **Rucks Pit Member** (late Gelasian Age Pleistocene). These members were described by Petuch and Roberts (2007: 139-145).

Stratotype. The type locality of Matson and Clapp's Nashua Formation is an exposure "one-quarter mile" south of Nashua, Putnam County, near the St. John's River. Here, 5 km of a "white shell marl" was exposed and was overlain unconformably by 1.5 m of white sand. Because this locality is now covered and lost, we here designate a locality adjacent to Nashua, that was described by Cooke and Mossom (1929: 225) " ... in a gully ... just south of the center of Sec. 28, T. 11 S., R. 26 E. (Putnam County, Florida), in the bluff east of the St. John's River." The stratotype exposure probably represents the upper beds of the Nashua, correlative with the Rucks Pit Member in southern Florida.

Age and Correlative Units. Based on the presence of widespread key index fossils such as the gastropods *Brachysycon willcoxi*, *Brachysycon amoenum*, *Scaphella brennmortoni*, and *Ximeniconus waccamawensis*, and the bivalves *Carolinapecten senescens*, *Noetia limula*, and *Conradostrea lawrencei*, the Nashua Formation is now known to span the entire Gelasian Pleistocene. The formation is exactly correlative with the Colerain Beach Member of the Chowan River Formation of Virginia and northern North Carolina, the Bear Bluff and Waccamaw Formations of southern North Carolina and northern South Carolina, and the Caloosahatchee Formation of southern Florida.

Fort Drum Member, Nashua Formation. In 1992, Thomas Scott, the Assistant State Geologist, discovered calcite spar geodes within a gray sandy limestone in the Rucks Pit at Fort Drum, Okeechobee County (also known as the Dickerson Corporation, Inc., Fort Drum Mine, or the Fort Drum Crystal Mine). Besides the presence of calcite geodes, the mine was of special interest to geologists in that it was continuously pumped, allowing for *in situ* collecting of fossil specimens and for direct study of the exposed geological formations. The Rucks Pit calcite geodes were highly unusual, because they had formed as aggregations of large dogtooth spar growing inside fossilized bivalves, mostly pairs of *Mercenaria permagna* (Figures 9.1 and 9.3), and large gastropods such as *Busycon* and *Sinistrofulgur* (Figure 9.6D). Nothing like this had ever been found before within the Everglades region. Scott also noted the presence of thick beds of fossils interbedded in a white calcarenite and observed that they were especially well preserved and contained a large number of species not normally found within the Everglades area. The senior author was brought to the mine by Dr. Scott in 1993 and soon discovered that the molluscan fauna contained not only

Figure 9.3 Close-up view of a bed of the large venerid bivalve, *Mercenaria permagna*, within an exposure of the Fort Drum Member, Nashua Formation in Rucks Pit, Fort Drum, Okeechobee County, Florida. Photo by Anton Oleinik.

new endemic species but also species originally described from the Waccamaw, James City, and Caloosahatchee Formations. These well-preserved mollusks were contained within the white, unconsolidated upper layer within the quarry (Figure 9.1), and this unit is undoubtedly the "white shell marl" of Matson and Clapp. The lower dark gray sandy limestone, however, had never before been reported or described.

Lithologic Description and Areal Distribution. As defined by Petuch and Roberts (2007: 137-139): "The Fort Drum Member consists of interbedded units of variable lithologies, including thick layers of a dark gray, very sandy (quartz) limestone or a dark gray, limy sandstone cemented by calcite microspar (arenaceous sparites), dense beds of the bivalve *Mercenaria* or busyconid gastropods in a matrix of gray or brown unconsolidated quartz sand (both seen in Figure 9.3 and 9.7), dark brown muddy sand filled with small broken shell bioclasts ("shell hash"), or lenses of light tan or cream-colored coquina (small shell fragments similar to the "shell hash" layers). Typically, the dark gray sandy limestone is extremely vuggy and moldic, often containing the partially dissolved remnants of *Mulinia* and *Noetia* valves. At certain localities, such as at Fort Drum, the main gray sandy limestone bed contains large molluscan bioclasts, primarily recalcified pairs of the bivalve *Mercenaria permagna.* These closed pairs are filled with large, honey-tan or golden-yellow crystals of iron-stained calcite" (shown here on Figures 9.4 and 9.5)

At the stratotype area in the Rucks Pit, the massive gray sandy limestone bed rests on a thick bed of large, perfectly preserved, closely packed single valves of *Mercenaria permagna* (apparently winnowed by tidal currents; Figure 9.3). This distinctive unconsolidated clam bed grades into the gray sandy limestone above and into the muddy "shell hash" layer below. The same massive gray arenaceous sparite-limy sandstone of the Fort Drum Member is also exposed within the Indrio Pit (Dickerson Corporation, Inc.) west of Fort Pierce, but there it lacks the *Mercenaria* geodes. In the Indrio Pit, the unconsolidated beds below the gray sandy limestone are not filled with *Mercenaria* clams but, instead, are replaced by beds of beautifully preserved busyconid whelks (several genera and species). North of Indian River County, the Fort Drum Member becomes predominantly a coarse quartz sand unit, dark brown or gray in color, with variable amounts of molluscan bioclasts. The areal extent of the Fort Drum Member is the same as that of the entire formation (Figure 9.2). In the area covered by this book, the member is present only in northeastern Okeechobee, St. Lucie, and northern Martin Counties. At the Rucks Pit, the Fort Drum Member averaged 6 m in thickness, whereas at the Indrio Pit it was over 7 m thick.

Stratotype. Petuch and Roberts (2007: 141) designated the stratotype of the Fort Drum Member as the lower beds on the southern wall of the Rucks Pit on Cemetery Road (NE 304th Street), 3 km east of U.S. Highway 441, and the town of Fort Drum, Okeechobee County, Florida. The type section is 6.2 m below land surface. The new member is named for the town of Fort Drum, Okeechobee County. The abandoned pit mine is now a lake on private property.

Age and Correlative Units. Based on the presence of widespread key index fossils, such as the gastropods *Brachysycon willcoxi* and *Ximeniconus waccamawensis* and the bivalves *Carolinapecten senescens* and *Mercenaria permagna*, the Fort Drum Member is now known to be of early Gelasian Pleistocene age. The member is correlative with the Colerain Beach Member of the Chowan River Formation of Virginia and northern North Carolina, the Bear Bluff Formation and lower beds of the Waccamaw Formation of southern North Carolina and northern South Carolina, and the Fordville and Fort Denaud Members of the Caloosahatchee Formation of southern Florida.

Fort Drum Index Fossils. Besides *Mercenaria* valves and busyconid whelks, very few other types of mollusks are known from the Fort Drum Member. This low diversity of fauna is probably the result of the harsh oceanographic conditions within the estuaries of the Nashua Lagoon System. These would have included cold winter water temperatures and fluctuating salinities. Most of the Fort Drum mollusks are wide-ranging species, being found from northern North Carolina to southern Florida. Some of these, and the unique Fort Drum calcite clam geodes, are shown here.

An exposure of the Fort Drum Member of the Nashua Formation at Rucks Pit, Fort Drum, Okeechobee County.

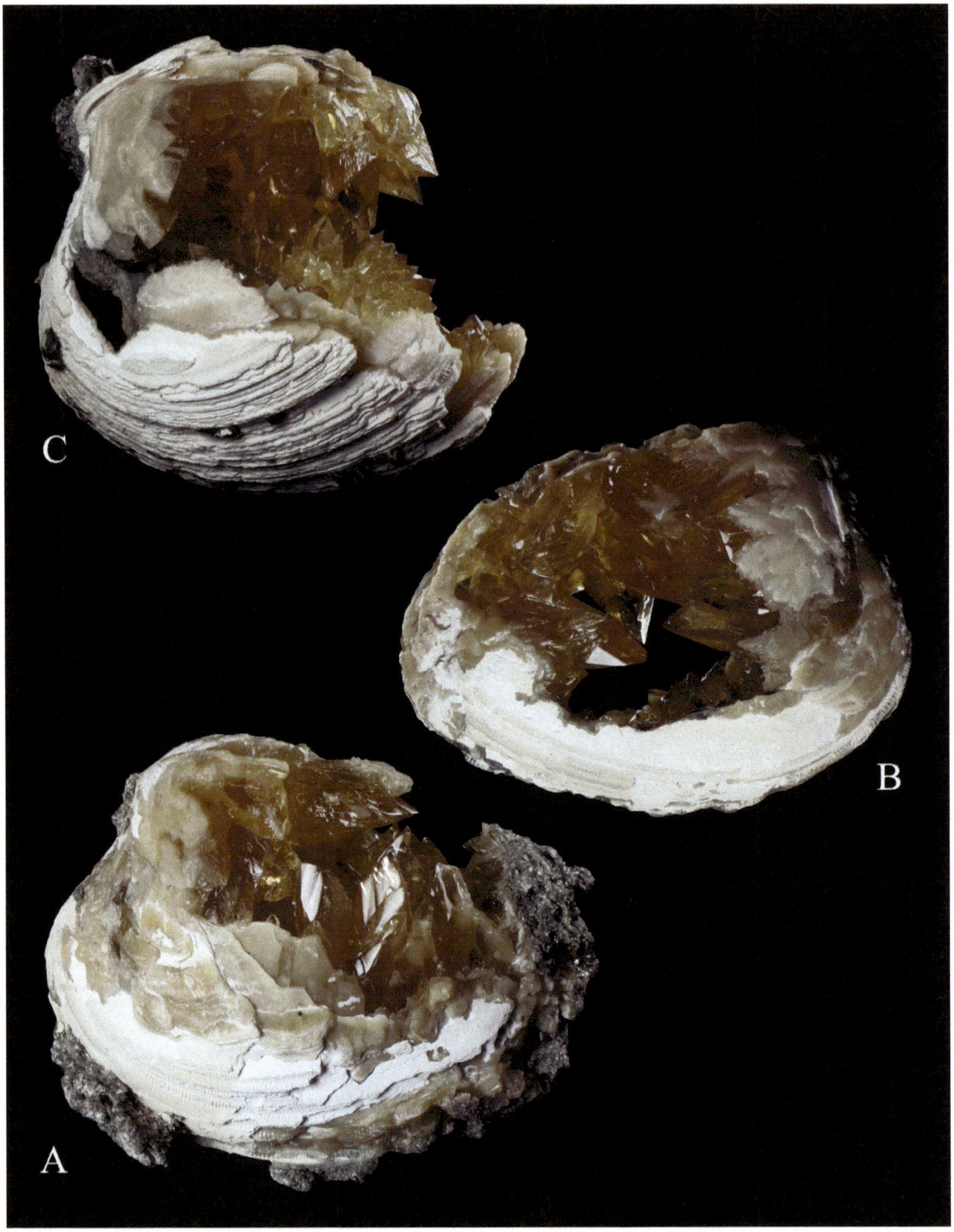

Figure 9.4 Venerid Bivalve Calcite Geodes of the Fort Drum Member, Nashua Formation.
A= *Mercenaria permagna* (Conrad, 1838), length 135 mm, with hinge line broken to reveal large dog tooth spar crystals; **B=** *Mercenaria permagna*, length 121 mm, with sides broken away to reveal large dogtooth spar crystals growing within interior; **C=** *Mercenaria permagna*, length 110 mm, half specimen broken to reveal a mixture of large dog tooth spar and smaller crystals.

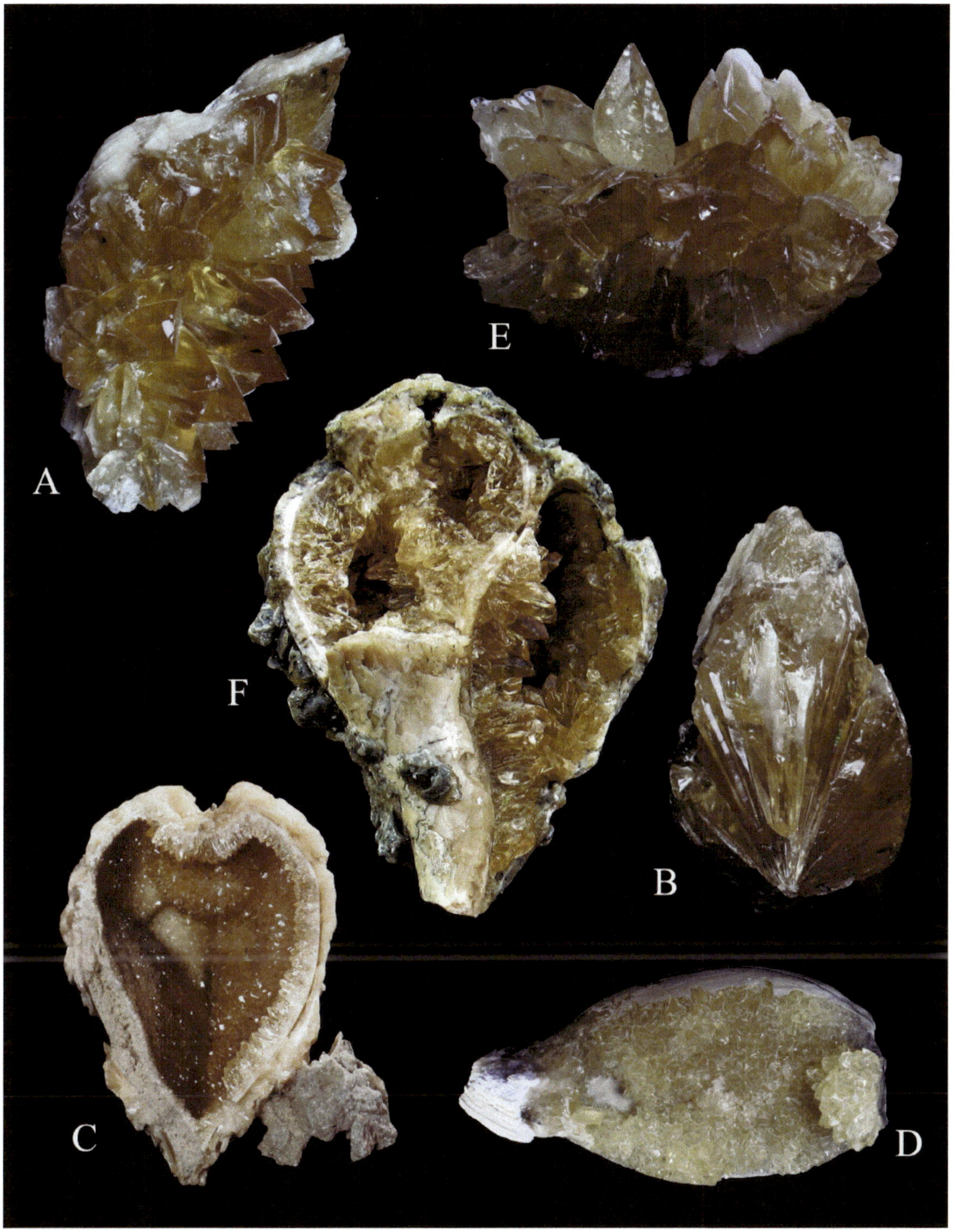

Figure 9.5 Calcite Crystals from the Fort Drum Member, Nashua Formation.
A= calcite flower that grew along the hinge area of a *Mercenaria permagna*, length 87 mm; **B**= calcite dogtooth spar crystals growing from the hinge line of a *Mercenaria permagna*, length 80 mm; **C**= druzy calcite crystals growing inside a heart-shaped, cross-sected *Mercenaria permagna*, length 99 mm; **D**= druzy filling along the hinge line of a *Mercenaria permagna*, length 97 mm; **E**= calcite flower that grew along the hinge area of a *Mercenaria permagna*, length 77 mm; **F**= *Busycon gilmorei* Petuch, 1994, calcite geode, length 135 mm.

Figure 9.6 Mollusks of the Fort Drum Member, Nashua Formation.
A= *Sinistrofulgur pamlico* Petuch, 1994, length 96 mm; **B**= *Busycon gilmorei* Petuch, 1994, length 165 mm; **C**= *Ximeniconus waccamawensis* (B. Smith, 1930), length 35 mm; **D**= *Sinistrofulgur pamlico* calcite geode, diameter 152 mm; **E**= *Noetia limula* (Conrad, 1832), calcite geode, length 48 mm; **F**= *Mercenaria permagna* (Conrad, 1832), length 142 mm, with edge chipping breakage due to a *Sinistrofulgur pamlico* predation attempt; **G**= *Brachysycon willcoxi* (Gardner, 1948), length 115 mm; **H**= *Americoliva rucksorum* (Petuch, 1994), length 42 mm.

Figure 9.7 View of an exposure of the Rucks Pit Member of the Nashua Formation in the Dickerson Aggregates, Inc. quarry, Fort Pierce, St. Lucie County, Florida. Here, a large aggregation of busyconid whelks has accumulated within a bed of *Mulinia lateralis* clams, along with the olivid *Americoliva roseae*. The majority of the whelks are *Busycotypus scotti* but other species such as *Busycon rucksorum*, *Brachysycon amoenum*, and *Sinistrofulgur yeehaw* are also present. This type of assemblage is typical of the Rucks Pit Member.

Rucks Pit Member, Nashua Formation. The excavations at Fort Drum in the early 1990s uncovered large exposures of a white, sandy unconsolidated lime mud (calcilutite) filled with small *Mulinia* bivalve shells. The molluscan assemblage was much richer and more diverse than that of the underlying gray sandy limestones and clam beds. Like the Fort Drum Member mollusks, most of the shells of the upper white layer were found to be of northern origin, primarily shared with the James City and Waccamaw Formations. The lithology of this unit and its molluscan fauna closely resemble the Nashua type section in Putnam County and may be correlative.

Lithologic Description and Areal Distribution. (From Petuch and Roberts, 2007: 142-144): "The Rucks Pit Member typically is composed of various mixtures of quartz sand and lime mud packed with mollusk shells (arenaceous bioclastic calcilutite). The quartz sand and fine particulate carbonates are white, cream-white, or light gray, weathering to white in exposed sections. In some facies, quartz sand dominates and contains a minor percentage of clay minerals, mostly sepiolite and montmorillonite. The percentages of clay minerals increase in the northern part of the member, whereas the carbonate fines dominate in the southern part. In other facies, the small estuarine bivalve *Mulinia lateralis* forms densely packed beds with only a small amount of calcilutites and quartz sand. Intercalated throughout the *Mulinia* beds are thin layers composed of beds of the scallop *Carolinapecten planicosta*, small bioherms of the oyster *Conradostrea lawrencei*, and larger bioherms of the ahermatypic coral *Septastrea* species. At the Indrio Pit west of Fort Pierce, the Rucks Pit Member is represented by a thick (2–3 m), white quartz sand layer with scattered beach-worn mollusk shells. The basal part of the *Mulinia* beds frequently indurates into a pale gray sandy coquina, but only forming thin scattered layers.

The areal extent of the Rucks Pit Member is the same as that of the entire formation (Figure 9.2). In the area covered by this book, the member is present only in northeastern Okeechobee, St. Lucie, and northern Martin Counties. At Fort Drum, the member averaged 3 m in thickness and at the Indrio Pit west of Fort Pierce, it was 3–4 m thick. The section given by Cooke (1945: 226) from the City Marl Pit near DeLand, Volusia County, contained an exposure of the Rucks Pit Member at 4 m below land surface. The lithology ("… creamy yellow sandy shell marl …") closely resembles that of the Rucks Pit stratotype.

Stratotype. We here designate the stratotype of the Rucks Pit Member as the southern wall of the Rucks Pit (also Dickerson Aggregates Corporation, Inc., Fort Drum Mine) on Cemetery Road (NE 304th Street), 3 km east of U.S. Highway 441 and the town of Fort Drum, Okeechobee County, Florida. The type section is 4 m below land surface. The new member is named for the Rucks Pit, Fort Drum, Okeechobee County, Florida. The abandoned quarry is now a lake on private land.

Age and Correlative Units. Based on the presence of several widespread key index fossils, such as the gastropods *Brachysycon amoenum*, *Pyruella bladenense*, and *Volutifusus halscotti* and the bivalves *Noetia limula*, *Conradostrea lawrencei*, and *Dinocardium hazeli*, the Rucks Pit Member is now known to date from the late Gelasian Pleistocene. The member is correlative with the upper beds of the James City Formation of Virginia and northern North Carolina, the upper beds of the Waccamaw Formation of southern North Carolina and northern South Carolina, and the Bee Branch and Ayers Landing Member of the Caloosahatchee Formation of southern Florida.

View of an exposure within the Rucks Pit, Fort Drum, Okeechobee County showing the *Mercenaria permagna* clam layer of the Fort Drum Member and the suprajacent cleaner sandy sediments of the Rucks Pit Member. This upper member of the Nashua Formation is being mined for road fill and is piled up in the background.

Rucks Pit Index Fossils. The Rucks Pit Member molluscan fauna, found within the Everglades area only in the northern Okeechobee Plain, comprises three separate components: an endemic component of restricted Nashua species, a component of northern James City and Waccamaw species, and a small component of tropical species shared with the Caloosahatchee Formation. Some of the more abundant and indicative species are shown on the following figures.

Figure 9.8 Index Fossils of the Rucks Pit Member, Nashua Formation.
A, B= *Siphocypraea (Floridacypraea) klatti* Petuch and Drolshagen, 2011, length 65 mm; **C, D**= *Pahayokea bruggemani* Petuch and Drolshagen, 2011, length 63 mm; **E**= *Volutifusus halscotti* Petuch, 1994, length 115 mm; **F**= *Septastrea* species, length 192 mm; **G**= *Carolinapecten planicosta* (Gardner, 1943), length 175 mm; **H**= *Cinctura rucksorum* (Petuch, 1994), length 115 mm; **I**= *Hexaplex* cf. *trippae* Petuch, 1994, length 110 mm; **J**= *Perplexiconus williamsonae* Petuch & Drolshagen, 2011, length 23 mm.

Figure 9.9 Mollusks of the Rucks Pit Member, Nashua Formation.
A= *Busycotypus scotti* Petuch, 1994, length 169 mm; **B**= *Busycon rucksorum* Petuch, 1994, length 215 mm; **C**= *Sinistrofulgur yeehaw* Petuch, 1994, length 152 mm; **D**= *Pyruella bladenense* (Gardner, 1948), length 142 mm; **E**= *Brachysycon amoenum* (Conrad, 1875), length 115 mm; **F**= *Stramonita* cf. *consul* (Chemnitz, 1795), length 95 mm; **G**= *Conradostrea lawrencei* Ward and Blackwelder, 1987, length 80 mm; **H**= *Stramonita* cf. *floridana* (Conrad, 1837), length 90 mm.

Close-up view of an exposure of the Rucks Pit Member of the Nashua Formation, in the Dickerson Aggregates Indrio Pit, west of Fort Pierce, St. Lucie County, Florida. Here, numerous specimens of the whelk *Busycon rucksorum* have accumulated within a lag deposit of an ancient tidal channel. Lag deposits and mass accumulations of mollusk shells are characteristic of the depositional patterns within the Gelasian Pleistocene Nashua Lagoon System.

CHAPTER 10. The Loxahatchee Subsea (Bermont Formation: Holey Land Member)

The mid-Pleistocene (1.8 million to 126,000 years B.P.) was a time of rapidly changing climates and sea levels. During some periods of time, such as 1 million years B.P., the climate approximated that seen in southern Florida today and sea level was close to that of the present (Figure 10.2). During other times, such as 780,000 years B.P., sea level plummeted to over 150 m below that of the present, and the entire Florida Platform became emergent. At that time, southern Florida probably had a climate similar to modern-day southern Georgia. These wildly fluctuating climates and sea levels had a devastating effect on the marine communities of the Everglades Pseudoatoll. This was manifested in the abrupt extinction of many prominent groups, along with the rapid evolution of completely new taxa (Petuch, 1995; 2004; Petuch and Roberts, 2007; Petuch, Myers, and Berschauer, 2015; Petuch, Berschauer, and Myers, 2018). The eustatic fluctuations of the middle Pleistocene came in two large pulses, the first starting at about 1.8 million years B.P. and the second around 781,000 years B.P. (Figure 10.2); these correspond with the initiations of the Chibanian and Ionian Ages. The first major transgression, at the beginning of the Calabrian Age, flooded the Everglades Basin with a new marine system, the **Loxahatchee Subsea** (named for the town of Loxahatchee, Palm Beach County; Petuch, 2004). Throughout Calabrian time, the narrow Loxahatchee Trough remained the only deep water area within the shrinking pseudoatoll. This enclosed, landlocked feature acted as a refugium for many tropical marine organisms from Caloosahatchee times. These survivors, mixed together with newly evolved organisms, gave the marine communities of the Loxahatchee Subsea a strange and unique appearance.

The complex pattern of sea level fluctuations during the middle Pleistocene manifested themselves in equally complex patterns of deposition within the Everglades area. The units that resulted from these complex patterns are here grouped together as a single large geologic unit, the Bermont Formation. Its lower member, the **Holey Land Member**, was deposited during the Calabrian Age, within the enclosed basin of the Loxahatchee Subsea. The middle member, the **Okeelanta Member**, was deposited within the environs of a Pleistocene paleo-lake during the low sea levels at the incipiency of the Chibanian Age. The upper member, the **Belle Glade Member**, was deposited during the major marine transgression during the subsequent Chibanian Age (both the Okeelanta and Belle Glade Members are discussed in detail in Chapter 11). The multiple lithofacies seen within the Holey Land Member of the Bermont Formation reflect the complex oceanographic conditions of the Loxahatchee Subsea and the surrounding Atlantic Ocean and Gulf of Mexico during the Calabrian Age. The middle Pleistocene Bermont Formation is the only Everglades stratigraphic unit to have been deposited within two different ages, two different Okeechobean subseas, and a giant fresh water lake.

Figure 10.1 Simulated satellite image (altitude: 100 miles) of the Floridian Peninsula during the Calabrian Pleistocene, showing the probable appearance of the **Loxahatchee Subsea** of the Okeechobean Sea. At this time, the Hendry Platform supported extensive growth of mangrove forests and sea grass beds and a new coral reef system, the **Monroe Reef Tract** formed off the Miami Archipelago of the southeastern edge of the subsea. The Monroe Reef Tract created the foundation for the modern Florida Keys.

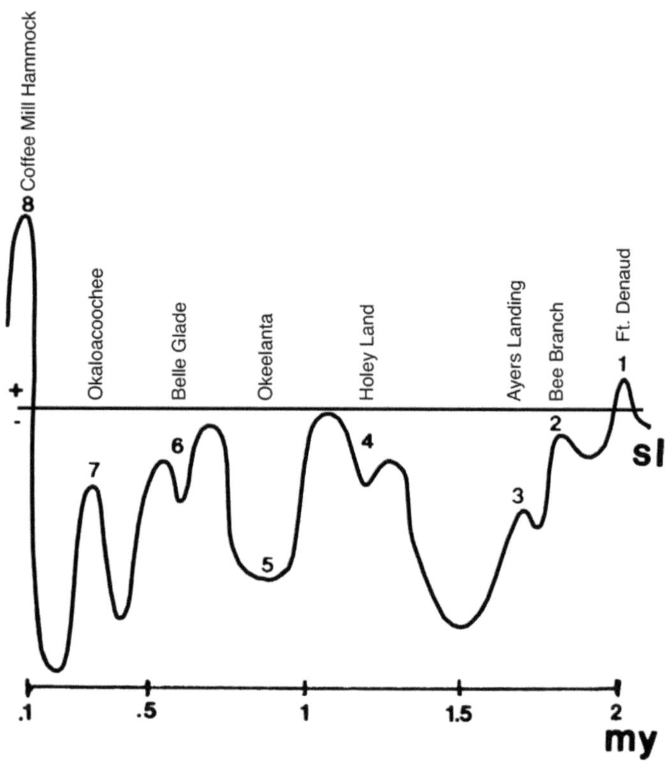

Figure 10.2 Sea level curve for the Pleistocene, showing major eustatic fluctuations. Present sea level (sl) is used as a reference standard. Time is in millions of years (my). Numbers correspond to the times of deposition of formations and members, during sea level highs. Formations: (1, 2, 3) = **Caloosahatchee Formation**; (4, 5, 6) = **Bermont Formation**; (7, 8) = **Fort Thompson Formation**. Members: 1 = **Fort Denaud Member** (Caloosahatchee Formation); 2 = **Bee Branch Member** (Caloosahatchee Formation); 3 = **Ayers Landing Member** (Caloosahatchee Formation); 4 = **Holey Land Member** (Bermont Formation); 5 = **Okeelanta Member** (Bermont Formation); 6 = **Belle Glade Member** (Bermont Formation); 7 = **Okaloacoochee Member** (Fort Thompson Formation); 8 = **Coffee Mill Hammock Member** (Fort Thompson Formation). Taken from the Oxygen Isotope Model of Krantz (1991) and from Vail *et al.* (1977).

Deposition within the Loxahatchee Subsea. For over 300,000 years, the Everglades area remained above sea level after the retreat of the Caloosahatchee Subsea. During this terrestrial interval, the Everglades contained a large freshwater lake (or possibly a series of smaller lakes), presaging present-day Lake Okeechobee and the Everglades. This early Calabrian paleolake, **Lake Immokalee** (named for the city of Immokalee, Collier County; Petuch, 2004: 228) is preserved as a thin (0.5 m) layer of freshwater limestone, filled with freshwater snails, on the surface of the Ayers Landing Member of the Caloosahatchee Formation (see Chapter 8). This paleolake limestone has been found at several localities, including Mule Pen Quarry, East Naples, Collier County, and the Holey Land Levee on the Miami Canal, Palm Beach County (observed *in situ* during water drawdowns). At around 1.75 million years B.P., the Loxahatchee Subsea began to flood Lake Immokalee and the Everglades Basin. This new subsea enlarged the geomorphologic features that had developed within the predecessor Caloosahatchee Subsea, producing a larger Immokalee Island that expanded to cover most of the Hendry Platform, leaving only a small strip of mangrove islands between it and Miccosukee Island. Similarly, the Kissimmee Embayment had filled with

Figure 10.3 View of the Palm Beach Aggregates Pit #6, Loxahatchee, Palm Beach County, Florida, showing exposed sections of the Bermont Formation. The alternating sandy limestone and unconsolidated carbonate sediments of the Holey Land Member can be seen to be overlain by the lighter-colored Belle Glade Member. Photo by Anton Oleinik.

mangrove jungles, and its northern half was completely emergent, filled with pine and oak forests. The eastern side of the pseudoatoll was also changing rapidly during the time of the Loxahatchee Subsea. Here, the Dade and Palm Beach reef tracts had evolved into a large chains of wide islands, covered with forests of pine trees and palmettos. On the western side of these islands, within the quiet, solar-heated waters of the Loxahatchee Trough, a series of large coral bioherms and fringing reefs had formed, being best developed along the southern end of the archipelago. Similarly, a long line of zonated coral reefs, the **Monroe Reef Tract** (named for Monroe County), was forming along the southeastern edge of the archipelago, growing seaward into the

Florida Straits. This reef tract marks the initiation of the deposition that would result in the formation of the modern Florida Keys. At the northern end of the Palm Beach Archipelago, a large chain of sand and mangrove islands was beginning to form, greatly narrowing the Loxahatchee Strait. Only the Caloosahatchee Strait remained deep and wide at this time, connecting the Loxahatchee Subsea to a broad, shallow area of coastal lagoons to the north.

The Bermont Formation, Okeechobee Group. In the early 1960s, when quarries were being dug in the vicinity of Belle Glade, Palm Beach County, Druid Wilson first recognized a lithologically distinct set of beds containing a new and unstudied molluscan fauna. Wilson never published his discoveries, but Axel Olsson later gave an overview of his research (Olsson *in* Olsson and Petit, 1964: 521–522), naming the set of beds "Unit A." This unnamed formation was also referred to as the "Glades Unit" and its molluscan fauna was reported on by Thomas McGinty (1970), as well as by Hoerle (1970; who gave a detailed species list, but with now-outdated taxonomy). Finally, in 1974, Jules DuBar named the Unit A-Glades Unit beds the **Bermont Formation**, for the Bermont quadrangle (near the hamlet of Bermont, Charlotte County). DuBar's new formation was described from his "Unit F" along the Charlotte County section of Shell Creek, a tributary of the Peace River. Based on the distribution of some key index fossils, DuBar was able to show that his Bermont Formation was widely distributed around southern Florida (DuBar, 1974: Table 8).

During the late 1980s and early 1990s, new quarry excavations and canal dredgings in Dade, Broward, and Palm Beach Counties exposed some of the thickest and best-developed beds of the Bermont Formation that had ever been seen previously. Closer examination of these exposures showed that DuBar's formation was actually far more complex than originally described, with a far greater range of lithologies and greater stratigraphic development. Likewise, the molluscan fauna of the Bermont Formation was also found to be far richer, containing more endemic taxa than originally thought (Petuch, 1988, 1989, 1991, 1994, 2004). Working in quarries in western Miami (Capeletti Brothers Pit #11) and southwestern Palm Beach County (Griffin Brothers Pit), the senior author (Petuch, 1991) also found that the Bermont Formation comprised at least two distinct members, each containing its own distinctive molluscan fauna. The Bermont Formation was also recognized as occurring in the Leisey Shell Corporation quarries in southern Hillsborough County and was reported on by Morgan and Hulbert, MacFadden, and Portell *et al.* in a special bulletin of the Florida Museum of Natural History (1995). Most recently, the paleoceanography and paleoecology of the Bermont Formation were described by the senior author (Petuch, 2004: 227–243). In the collection of review papers entitled *The Geology of Florida* (Randazzo and Jones, 1997), the Bermont Formation is missing from the generalized stratigraphic column shown on the inside covers. Jones (115–116, Figure 7.22), however, later discusses the Bermont Formation within the same book and illustrates Bermont index fossils (from the "Pleistocene Bermont Formation of southern Florida").

Lithologic Description of the Bermont Formation. Condensed here is an extract from DuBar (1974: 221–222): "The formation has been traced along Shell Creek for a distance of about seven miles. The unit typically is a gray, unconsolidated sandy shell marl. The sand fraction is composed of clear, subrounded, medium-grained quartz with a low percentage of admixed rounded, lustrous black phosphorite grains. Fossils are abundant and well preserved, and assemblages are quantitatively and qualitatively homogeneous throughout the area … An apparent facies of the Bermont Formation occurs in Palm Beach County along the southeast margin of Lake Okeechobee between Belle Glade and Port Mayaca … the Belle Glade facies is a gray shell marl." Morgan and Hulbert (1995: 21) described the Bermont beds exposed in the Leisey Shell pits as "1.0 m, dark, sandy mud; freshwater shells; bones" (lower bed) and "2.3 m, cream to bluish sandy, massive shell bed" (upper bed). The formation description is here expanded to include thick beds of heavily indurated fossiliferous limestones (biomicrudites and biocrudites, both marine and freshwater), highly arenaceous limestones (and limy sandstones), often containing petrified wood, unconsolidated yellow-tan quartz sands with scattered molluscan bioclasts, and lime muds (calcilutites) composed mainly of charaphyte phytoliths and packed with freshwater snails. The freshwater and marine limestone units are typically colored a light gray, whereas the sandy limestones are colored either golden-yellow or cream. Some limy sandstones are gray-brown to nearly black with dark gray or black silicified mangrove wood or dark gray silicified Turtle Grass (*Thalassia*) root casts.

With an expanded lithologic description and a larger and better-defined set of index fossils, the Bermont Formation is now known to be much more widely distributed than originally thought. As shown here in Figures 10.5 and 11.6 (for the two most widespread members), the Bermont Formation extends from southern Hillsborough County south to Monroe County and from southern Highlands and Okeechobee Counties south to Dade County. DuBar (1974: 221) states that the Bermont Formation extends northward up both coasts of Florida, to Levy County in the west and to Putnam County in the east. On the west coast, the beds north of southern Tampa Bay appear to be younger units, equivalent to the Fort Thompson Formation and do not belong in the Bermont Formation. Likewise, on the east coast, the beds north of Martin County appear to belong to the older Nashua Formation and also do not belong in the Bermont. Beach deposits above these Nashua beds, containing mostly quartz sand, may be equivalent to the Bermont. At the Rucks Pit in Fort Drum, Okeechobee County, the Bermont equivalent is represented by a thin (0.8 m) reddish-brown peaty sand layer, devoid of fossils, between the top of the Rucks Pit Member of the Nashua Formation and the base of the shell beds of the Coffee Mill Hammock Member of the Fort Thompson Formation.

Throughout its range, the Bermont is variable in thickness, being 3 m thick at the stratotype on Shell Creek, 10 m thick at Belle Glade, Palm Beach County, 4.5 m thick at the Leisey Shell pits, Hillsborough County, 12 m thick at Loxahatchee, Palm Beach County (Palm Beach Aggregates, Inc., quarries), and 15 m thick at the Capeletti

Figure 10.4 View of an exposure of the entire Bermont Formation in the Palm Beach Aggregates Pit #6, Loxahatchee, Palm Beach County, Florida. Here, the eroded gray sandy limestones and shell stringers of the Holey Land Member (unconsolidated facies) can be seen at the base of the section, while a thin bed of fresh water calcarenite and snails can be seen in the middle of the section. The yellow-colored Belle Glade Member (referred to locally as "Golden Rock") can be seen to comprise the upper third of the section. Photo by Anton Oleinik.

Brothers Pit in Miami, Dade County. The surface contours are also variable, being 6m below surface at the Leisey Shell pits, 5 m below Palm Beach Aggregates quarries, and 4 m below surface at the Griffin Brothers Pit in the Holey Land area. In the Everglades Agricultural Area (EAA), south of State Road 80 and between Belle Glade and the 20-Mile Bend at Loxahatchee, the surface of the Bermont undulates, ranging from 1.5 m to 5 m below land surface. These undulations are oriented in a northeast–southwest trend, are regularly spaced, and probably represent elongated sand bars and tidal channels that formed in the Loxahatchee Subsea. Within the Everglades Basin, the Bermont is the principal near-surface geologic formation.

Stratotype. From DuBar (1974: 221): "… the name Bermont Formation is informally proposed here for the uppermost fossiliferous marine sands exposed along Shell Creek in the Bermont quadrangle, Charlotte County." The Bermont quadrangle is centered on the hamlet of Bermont, at the intersection of State Road 31 and State Road 74 (Bermont Road; the only building at this intersection, the old Bermont General Store, burned down in 1998). Shell Creek is a small tributary of the Peace River, flowing into it just north of its confluence with Charlotte Harbor. The cypress heads and marshlands around Bermont house the headwaters of Shell Creek; DuBar (1974: 221) stated that the "formation has been traced along Shell Creek for a distance of about seven miles." This is considered the stratotype area.

Age and Correlative Units. Based on paleomagnetic studies done at the Leisey Pit in Hillsborough County (MacFadden, 1995), the basal beds (Holey Land Member) date from around 1.4 to 1.6 million years B.P. (Calabrian Age). Using uranium isotope ratio dating, the upper Bermont beds at Belle Glade (Belle Glade Member) were shown to date from approximately 750,000 to 800,000 years B.P. (Chibanian Age) (Lyons, 1991: 156–157). From these data, the Bermont Formation is now known to span 1.67 million years and is correlative with the Flanners Beach Formation of the Carolinas.

Bermont Index Fossils. Because the formation represents the deposition of two subseas and a paleolake, three sets of index fossils are recognized. Only a few species occurred in both the Loxahatchee and Belle Glade Subseas and only a few are found in both the Holey Land and Belle Glade Members. Some of these long-lived taxa include *Fasciolaria okeechobeensis*, *Aristofusus watermani*, *Vasum floridanum*, *Scaphella capelettii*, and *Strombus evergladesensis*. The faunas of the Holey Land and Belle Glade Members, for the most part, contain very different-looking molluscan assemblages, each with a very high percentage of endemism. The freshwater molluscan fauna of Pleistocene Lake Okeelanta (Okeelanta Member) is one of the most bizarre ever discovered and easily compares with the African Rift Valley lakes for diversity and species radiations. The index fossils of the Okeelanta and Belle Glade Members will be discussed and illustrated in Chapter 11.

Holey Land Member, Bermont Formation. In 1983, the senior author and his wife Linda, while collecting at the Griffin Brothers quarries on the eastern edge of the Holey Land Wildlife Management Area of Palm Beach County, discovered a previously unknown lithofacies of the Bermont Formation. This unit, which was later informally named the "Holey Land Limestone" (Petuch, 1990), was found to contain a highly endemic molluscan fauna, with many unusual taxa that had never been reported before. Besides containing classic Bermont index species such as *Miltha carmenae* and *Fasciolaria okeechobeensis*, the Holey Land Limestone also contained such oddities as the giant red-spotted olive shell *Lindoliva* (the largest-known member of the Olividae) and the siphonate melongenid subgenus *Miccosukea*. In 1986, the senior author encountered a Holey Land molluscan fauna in the Capeletti Brothers Pit #11, west of Miami, complete with *Lindoliva* and *Miccosukea* species. Unlike the Holey Land Limestone, these fossils were embedded in an unconsolidated sandy calcilutite (lime mud) and were exceptionally well preserved, retaining their original gloss and color patterns. (As a note to readers interested in Florida history; the area adjacent to the Griffin quarries was given the name "Holey Land" because it is peppered with shallow bomb craters, the result of having been an aerial bombing practice range during World War II. The bombers flew out of the Boca Raton Air Field, which is now Florida Atlantic where the senior author taught for 30 years; the old landing strips are now buried beneath the university stadium).

On the northwestern side of the Holey Land, a new canal and levee system was excavated and built in 1999, giving access to new exposures of the Holey Land Limestone. Some of the canal areas were diked off (such as at the Talisman Canal) and later dewatered, allowing for the first *in situ* collecting of Bermont strata anywhere in the central Everglades. Here, as in the Griffin pits on the eastern side of the Holey Land, the thick limestone unit was seen to be overlain by a layer of unconsolidated sediments filled with classic Bermont fossils such as those seen at Belle Glade. A layer of freshwater calcarenite, filled with planorbid snails, was also present between the limestone and the Belle Glade-type unit. These canal sections substantiated the senior author's (Petuch, 1988, 1990) previous idea that the Bermont was composed of at least two separate members and was not a single unit. In 1996, the senior author discovered another set of Bermont exposures in the Palm Beach Aggregates, Inc., quarries at Loxahatchee, Palm Beach County, west of West Palm Beach (previously known as the GKK Pit) (Figures 10.3 and 10.4). At this locality, all three main units of the Bermont Formation were well developed and the lower set of beds was found to contain the same molluscan fauna as that of the Holey Land Limestone. The genus *Lindoliva* was present here, but rare, and other normally-uncommon lower Bermont taxa were abundant. This indicates that a different set of depositional environments and ecosystems were present in the Loxahatchee area during Calabrian time. The lower beds of the Bermont at Palm Beach Aggregates also differed lithologically from those in Miami and the Holey Land area, consisting of fossiliferous sandy limestone units alternating with thin unconsolidated shell beds in a matrix of sandy lime mud. This distinctive unit was informally referred to as the "Loxahatchee Unit" by the senior author (Petuch, 2004:18). These beds were lithologically intermediate between the unconsolidated sandy calcilutites of the Capeletti Brothers Pit and the massive indurated fossiliferous limestone of the Griffin Brothers Pit, forming a continuum.

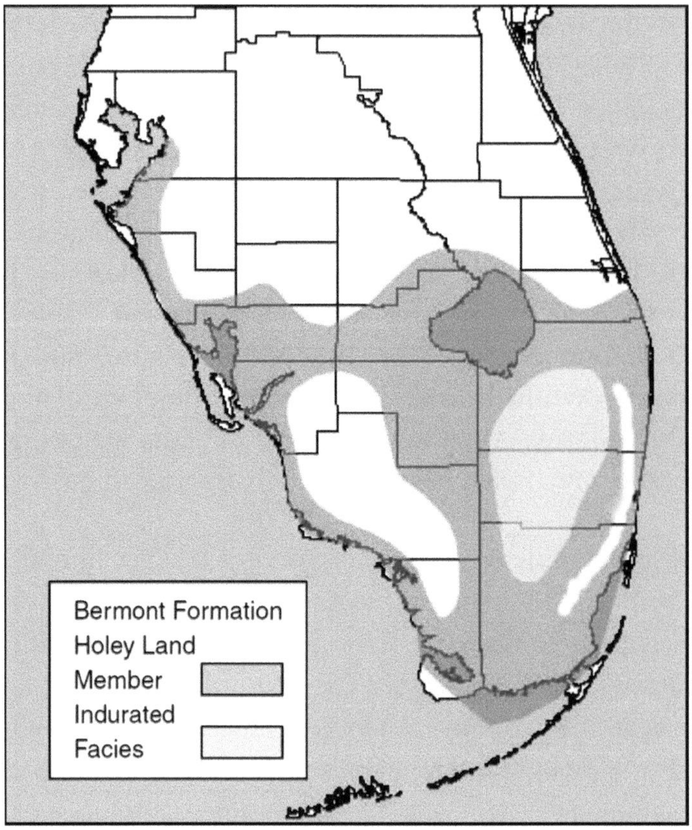

Figure 10.5 Areal extent of the Holey Land Member, Bermont Formation, in southern Florida.

Lithologic Description and Areal Distribution. Petuch and Roberts (2007: 154-157) gave a formal expanded definition of the senior author's (Petuch, 1990) original informal description for the Holey Land and its lateral equivalents; "The Holey Land Member typically consists of sandy (quartz), highly fossiliferous limestones (arenaceous biomicrudites and arenaceous crudites), often occurring in thick, heavily indurated units, beds of unconsolidated sandy (quartz) lime muds (arenaceous calcilutites) with densely packed molluscan bioclasts (primarily bivalves), and semifriable limy sandstones with high concentrations of organic matter (peats), containing silicified mangrove wood and Turtle Grass rhizomes. The fossiliferous limestones (Figure 6.4) are most commonly light gray to cream-gray in color, with white molluscan bioclasts (such as at the Holey Land stratotype), but can range in color from dark gray to grayish-brown (such as at the Palm Beach Aggregates quarries). The organic-rich, fine-particulate limy sandstones are generally dark brown, but can vary from brownish gray to black. Petrified mangrove wood, if present, is usually colored a dark gray or grayish-brown. The unconsolidated sandy lime muds (arenaceous calcilutites and calcarenites) vary in color from yellow tan to light blue-gray (at the Capeletti Brothers Pit), to bluish (at the Leisey Pit), to dark brownish-gray (at Palm Beach Aggregates quarries). The dense aggregations of molluscan fossils contained in these unconsolidated units vary from a distinctive blue-gray color (at Capeletti Brothers Pit #11) to bluish-white and pale tan (at Palm Beach Aggregates)."

Under the northern and central Everglades of southwestern Palm Beach County, western Broward County, and northwestern Dade County (Figure 10.5), the Holey Land is completely indurated into a very hard (possibly the hardest limestone in southern Florida), massive fossiliferous limestone unit. To the north and east of the indurated area in central Palm Beach County, the Holey Land contains a mixed lithology, with alternating thin beds of fossiliferous limestones and unconsolidated shell beds in a fine quartz sand and calcilutite matrix (the "Loxahatchee Unit" of Petuch, 1994). Elsewhere, from Hillsborough County to Dade County, the Holey Land is composed almost completely of unconsolidated sandy calcarenites and calcilutites, often containing dense shell beds. The shell beds are most commonly composed of bivalves, primarily a subspecies of the small venerid *Chione elevata* (previously referred to as *C. cancellata*, a related species from the West Indies), but may be composed of strombid gastropods (*Strombus*) or *Pyrazisinus* potamidid gastropods. Some facies, such as those found in the Capeletti and Palm Beach Aggregates quarries, contain rich assemblages of corals, often with over 20 species being present (Petuch, 2004: 235–236). At the Griffin Brothers Pit, the bedding plane surface of the massive limestone was frequently covered with a pavement of *Manicina areolata* rose corals and *Clypeaster rosaceus* echinoids. The Holey Land Member has the same areal distribution as does the entire Bermont Formation, extending from southern Hillsborough County to Monroe County and from southern Okeechobee County to Dade County. The member is absent under the Immokalee Rise and Atlantic Coastal Ridge, as these were islands during the time of deposition. Here, the Holey Land is replaced with undifferentiated sands. Throughout its range, the Holey Land Member varies in thickness from 3.5 m at the Leisey Shell pit, to over 6 m at the Griffin Brothers Pit, to 10 m at Palm Beach Aggregates, to over 15 m at the Capeletti Brothers Pit #11. The heavily indurated area of the Holey Land (Figure 10.5) roughly corresponds to the basin of Lake Okeelanta (described in Chapter 11). These massive limestones may have formed diagenetically through exposure in the deeper water areas of the paleolake. Here, the Holey Land sediments would have constituted the exposed floor of Lake Okeelanta, making them vulnerable to infiltration and recrystalization. Holey Land sediments covered by the peripheral marshland sediments of Lake Okeelanta, on the other hand, were protected from extensive diagenesis and remained unconsolidated.

Stratotype. Retaining the name and original type locality, Petuch and Roberts (2007) designated the stratotype of the Holey Land Member to be the western side of the Griffin Brothers Pit #2, directly adjacent to the Holey Land Wildlife Management Area, on the power line access road, 7 km west of Deem City, at the intersection of Highway 27, Palm Beach–Broward County line, Palm Beach County, Florida. The Griffin pits are now flooded and have been incorporated into the Everglades Restoration Plan, as part of the filter marsh system. The Holey Land stratotype lies 2.5 m below mean water surface and can be accessed by scuba diving. (Deem City, once a hamburger shop and trailer, is now gone).

Age and Correlative Units. Based on uranium isotope dating (discussed earlier in this chapter) and on the presence of key widespread index fossils such as the bivalve *Noetia platyura*, the Holey Land Member is now known to date from the Calabrian Age of the early middle Pleistocene. The member is correlative with the lower beds of the Flanners Beach Formation of the Carolinas.

Holey Land Index Fossils. The Holey Land Member, including all its facies, contains a rich and highly endemic molluscan fauna. These are made up of three main components: widespread species found throughout the full extent of Holey Land deposition; species found only within the central area of the pseudoatoll (Loxahatchee Trough); and relictual species that represent Kissimmean genera that survived into the Calabrian Pleistocene. Some of these relictual molluscan genera, surviving into Calabrian time only inside the Loxahatchee Subsea, include the gastropods *Pyruella*, *Calusaconus*, *Pseudozonaria*, *Malea*, and *Jenneria*, and the bivalves *Carolinapecten*, *Armamiltha*, and *Conradostrea*. All these relictual genera, with the exception of *Calusaconus*, were extinct or extirpated in Florida by the end of Holey Land time, and do not occur in the younger Belle Glade Member. Characteristic Holey Land index fossils are shown in the following iconography.

Close-up view of typical fossiliferous limestone from the Holey Land Member, showing *in situ* specimens of *Cinctura capelettii*, *Carolinapecten jamieae*, *Miltha carmenae*, and fragments of *Manicina* corals. From the Griffin Brothers Pit, Holey Land Wildlife Management Area, Palm Beach County, Florida.

Figure 10.6 Cypraeoidean Gastropods of the Holey Land Member, Bermont Formation. **A, B=** *Macrocypraea (Lorenzicypraea) spengleri* (Petuch, 1990), length 130 mm; **C, D=** *Macrocypraea joanneae* Petuch, 2004, length 54 mm; **E, F=** *Pseudozonaria portelli* (Petuch, 1990), length 23 mm; **G, H=** *Luria voleki* Petuch, 2004, length length 37 mm. **I=** *Cyphoma tomeui* Petuch and Drolshagen, 2011, length 26 mm; **J=** *Cyphoma voleki* Petuch and Drolshagen, 2011, length 21 mm; **K=** *Simnialena donovanae* Petuch, 1994, length 17 mm.

Figure 10.7 Gastropods of the Holey Land Member, Bermont Formation.
A= *Jenneria hepleri* Olsson, 1967, length 22 mm (only found in the lowermost bed of the Holey Land Member; the holotype is lost, this specimen is the potential neotype); **B=** *Jenneria loxahatchiensis* M. Smith, 1936, length 29 mm (upper beds of the Holey Land Member); **C=** *Pusula dadeensis* Petuch, 1994, length 20 mm; **D=** *Pusula lindajoyceae* Petuch, 1994, length 22 mm; **E=** *Semicassis loxahatcheensis* Petuch, 1994, length 61 mm; **F=** *Melongena (Rexmela) griffini* Petuch, 1994, length 35 mm; **G, H=** *Malea petiti* Petuch, 1989, length 122 mm; **I=** *Triplofusus papillosus* unnamed subspecies, length 280 mm.

Figure 10.8 Helmet Shells of the Holey Land Member, Bermont Formation.
A, B= *Cassis rasae* Petuch and Berschauer, 2021, length 145 mm; **C, D**= *Cassis viliusi* Petuch and Berschauer, 2021, length 165 mm; **E, F**= *Cassis schnireli* Petuch, 1994, length 183 mm; **G, H**= *Cassis jameshoubricki* Petuch, 2004, length 192 mm. The Holey Land Member contains the largest fauna of fossil *Cassis* species known from anywhere on Earth.

Figure 10.9 Mollusks of the Holey Land Member, Bermont Formation.
A= *Americoliva wendyae* (Petuch, 1994), length 50 mm; **B=** *Pyruella tomeui* Petuch, 2004, length 82 mm; **C=** *Sinistrofulgur holeylandicus* Petuch, 1994, length 248 mm (differs from other *Sinsistrofulgur* species in having a fine-threaded shell sculpture); **D=** *Fulguropsis capelettii* Petuch, 1994, length 122 mm; **E=** *Miltha carmenae* H. Vokes, 1969, length 82 mm; **F=** *Carolinapecten jamieae* Petuch, 2004, length 94 mm; **G=** *Vokesimurex anniae* (M. Smith, 1940), length 41 mm; **H=** *Vokesimurex diegelae* Petuch, 1994, length 36 mm; **I=** *Polygona maxwelli* (Pilsbry, 1930), length 47 mm; **J=** *Fasciolaria okeechobeensis* Tucker and Wilson, 1932, length 240 mm.

Figure 10.10 Close-up of Oleinik's Conch Shell, *Macrostrombus oleiniki*, exposed within a sandy unconsolidated facies of the Holey Land Member, Bermont Formation, in the Bergeron Star Ranch Pit, south of South Bay, Palm Beach County, Florida. Photo by Anton Oleinik.

Figure 10.11 Strombid Gastropods of the Holey Land Member, Bermont Formation.
A= *Strombus erici* Petuch, 1994, length 95 mm; **B=** *Strombus capelettii* Petuch, 1994, length 69 mm; **C=** *Lobatus wilsonorum* (Petuch, 1994), length 54 mm; **D=** *Macrostrombus diegelae* (Petuch, 1991), length 94 mm; **E=** *Macrostrombus whicheri* Petuch and Drolshagen, 2011, length 148 mm; **F=** *Macrostrombus oleiniki* Petuch and Drolshagen, 2011, length 221 mm; **G=** *Titanostrombus williamsi* (Olsson and Petit, 1964), length 265 mm; **H=** *Macrostrombus tomeui* Petuch and Drolshagen, 2011, length 146 mm.

Figure 10.12 Mollusks of the Holey Land Member, Bermont Formation.
A= *Scaphella capelettii* Petuch, 1994, length 89 mm; **B**= *Scaphella seminole* Petuch, 1994, length 52 mm; **C**= *Lindoliva spengleri* Petuch, 1988, length 110 mm; **D**= *Americoliva ryani* Petuch, 1994, length 58 mm; **E**= *Americoliva smithorum* Petuch, 1994, length 47 mm; **F**= *Porphyria gravesae* (Petuch, 1994), length 57 mm; **G**= *Vasum floridanum* McGinty, 1940, length 117 mm; **H**= *Macrostrombus scotti* (Petuch, 1994), length 183 mm; **I**= *Lindoliva griffini* Petuch, 1988, length 73 mm (confined to the lowest beds of the Holey Land Member); **J**= *Aliger pahaokee* (Petuch, 1994), length 202 mm.

Figure 10.13 Gastropods of the Holey Land Member, Bermont Formation.
A= *Lindaconus swearingeni* Petuch and Drolshagen, 2011, length 59 mm; **B=** *Lindaconus spengleri* (Petuch, 1991), length 79 mm; **C=** *Gradiconus loxahatcheensis* (Petuch, 1994), length 58 mm; **D=** *Calusaconus tomeui* Petuch, 2004, length 28 mm; **E=** *Gradiconus capelettii* (Petuch, 1994), length 42 mm; **F=** *Perplexiconus alligator* Petuch and Drolshagen, 2011, length 32 mm; **G=** *Purpuriconus erici* Petuch and Drolshagen, 2011, length 22 mm; **H=** *Jaspidiconus hyshugari* (Petuch, 1994), length 32 mm; **I=** *Macrostrombus holeylandicus* (Petuch, 1994), length 194 mm; **J=** *Macrostrombus mayacensis* (Tucker and Wilson, 1933), length 152 mm.

Figure 10.14 Mangrove-Associated Index Fossils of the Holey Land Member, Bermont Formation. **A**= *Avicennia germinans* Black Mangrove agatized branch ("petrified wood"), length 320 mm; **B**= *Avicennia germinans* Black Mangrove agatized branch ("petrified wood"), length 178 mm; **C**= *Pyrazisinus roseae* Petuch, 1991, length 58 mm; **D**= *Pyrazisinus turriculus* Petuch, 1994, length 62 mm; **E**= *Pyrazisinus palmbeachensis* Petuch, 1991, length 42 mm; **F**= *Melongena lindae* Petuch, 1994, length 68 mm; **G**= *Cerion petuchi* Harasewych, 2012, holotype, length 36 mm (lived on grasses along the shoreline); **H**= *Miccosukea cynthiae* (Petuch, 1990), length 61 mm.

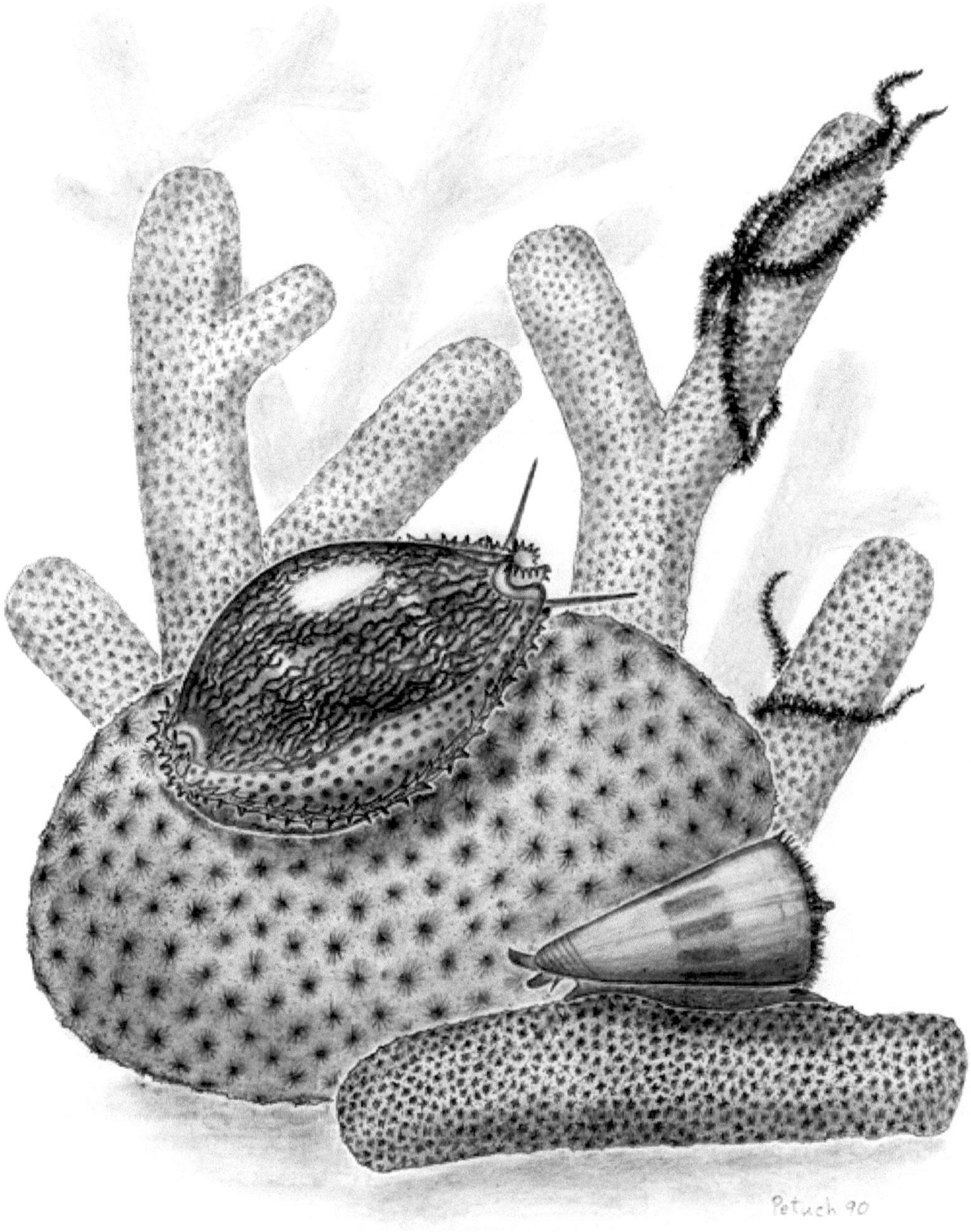

An artistic rendering of a *Porites* coral thicket that grew along the Palm Beach Archipelago of the Loxahatchee Subsea during the Calabrian Pleistocene. Here, the small cowrie, *Pseudozonaria portelli*, can be seen crawling on a *Siderastrea* coral head while a specimen of the cone shell, *Cariboconus griffini*, crawls on a fragment of *Porites* coral in the foreground. This assemblage is typical of the coral reef facies of the Holey Land Member of the Bermont Formation. Taken from Petuch, 1992.

Chapter 11. Lake Okeelanta and the Belle Glade Subsea (Bermont Formation: Okeelanta and Belle Glade Members)

From 1980 until 1994, a section of the North New River Canal, along State Highway 27, 3–5 km south of South Bay, Palm Beach County (adjacent to the Okeelanta Corp. sugar mill property), was regularly dredged and widened. During these excavations, large blocks of a highly fossiliferous pale orangish-tan calcarenite were brought up by the draglines and tossed onto the top of the canal berm (capped with a dirt access road). The senior author visited this site on a continuous basis over the time of the excavations, and discovered that these "marl" blocks were filled with a highly unusual freshwater molluscan fauna. Of particular interest was a large number of species of planorbid gastropods (genera *Planorbella* and *Seminolina*), including bizarre elongated and uncoiled forms that had never before been reported. These distinctive index fossils were later found at other localities along the North New River Canal and Highway 27, from 5 to 20 km south of the Okeelanta site. After extensive field collecting over several years, the blocks of "marl" and the abundant fresh water snails were found to have been deposited within a large freshwater paleolake, which was later named **Lake Okeelanta** after the site of the first exposures (Petuch, 1997: 311-318; Petuch and Roberts, 2007). This Everglades paleolake formed during the sea level low at the beginning of the Chibanian Age and at its maximum development, during this emergent time, was a deep (at least 20 m) holomictic lake that encompassed an area approximately four times the size of Recent Lake Okeechobee (Figure 11.1).

In the walls of the dewatered canals exposed during the construction of the Holey Land levee (discussed in the previous chapter), the Lake Okeelanta freshwater calcarenite was found to be a much more complicated and thicker unit, also containing limestone layers, beds of freshwater clams, and charaphyte phytolith lime muds (phytolithic calcilutites). At the Holey Land levee site, these freshwater beds were subjacent to a sandy calcarenite filled with typical Belle Glade marine fossils. This same depositional pattern was seen during dragline dredgings in the Griffin Brothers Pit. The limestone units at both localities contained either densely packed assemblages of the tiny hydrobioid gastropod *Fontigens*, small *Seminolina* planorbid gastropods, or beds of unnamed species of *Corbicula* and freshwater *Rangia* clams. The Okeelanta unit is also present in the Palm Beach Aggregates quarries at Loxahatchee, but here it is poorly developed and forms a thin bed between the Holey Land and Belle Glade Members (see Figure 10.4, Chapter 10). At this locality, *Seminolina* snails are outnumbered by a large discoidal *Planorbella* species, *Viviparus* snails, *Stenophysa* species, and *Pomacea* apple snails (all illustrated in the following sections). Occupying the basin of the Okeechobean Sea, Lake Okeelanta formed in a rich carbonate sedimentary environment, and its water was saturated with dissolved calcium carbonate. This saturation is reflected in the thick layers of charaphyte phytolith lime muds that were produced by calcareous green algae, such as *Chara*, which thrive in this type of water chemistry. Lake Okeelanta was analogous to

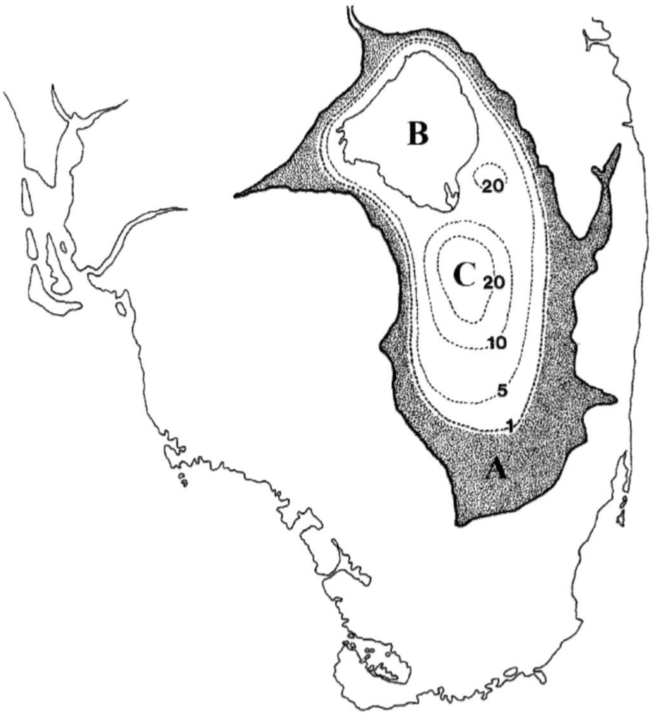

Figure 11.1 Map of Pleistocene Lake Okeelanta (taken from Petuch, 1988:129). **A= Marshland Zone** (stippled shading); **B= Caraphyte Zone**; **C= *Corbicula* Zone**. Depth contours in meters below lake surface.

an inland sea, producing the richest and most diverse lacustrine molluscan fauna known from anywhere in North America.

Based on the distribution of the Okeelantan molluscan assemblages and on the ecology of living related forms, three broad, interfingering ecological zones can be delineated within the paleolake boundaries: (1) a peripheral **Marshland Zone**, dominated by emergent vegetation and containing a molluscan fauna made up of *Viviparus*, *Stenophysa*, *Pomacea*, and large discoidal *Planorbella* snails; (2) a **Charaphyte Zone**, comprising shallow, open bottom areas that supported dense beds of the green calcareous alga *Chara*; and (3) a ***Corbicula* Zone**, with dense beds of *Corbicula* and *Rangia* bivalves living on open bottoms in deeper water. The bizarre, elongated *Seminolina* species shown in Figure 11.3 lived in the *Corbicula* Zone, and this assemblage represents the single most unusual freshwater community ever found in Florida. Each of the ecological zones was also associated with distinct types of deposition, ranging from muddy sands with high levels of organic matter to pure, clean carbonate muds (see Petuch, 1997 for the paleolimnology of Pleistocene Lake Okeelanta).

Okeelanta Member, Bermont Formation

Lithologic Description and Areal Distribution. The Okeelanta Member typically consists of interbedded layers of freshwater calcarenites, generally highly fossiliferous. These range from unconsolidated lime muds (calcilutites) packed with planorbid gastropod or corbiculid and mactrid bivalve bioclasts, charaphyte phytolithic calcarenites with planorbid gastropod bioclasts, sandy (quartz) lime muds with high concentrations of organic matter (organic-rich arenaceous calcilutites) and with large molluscan bioclasts, mostly *Viviparus*, *Pomacea*, and discoidal planorbids, and thin fine-grained limestone units (biomicrudites) with abundant molluscan bioclasts. The calcilutites vary in color from pale orange-tan, to light brown, light gray, or pure

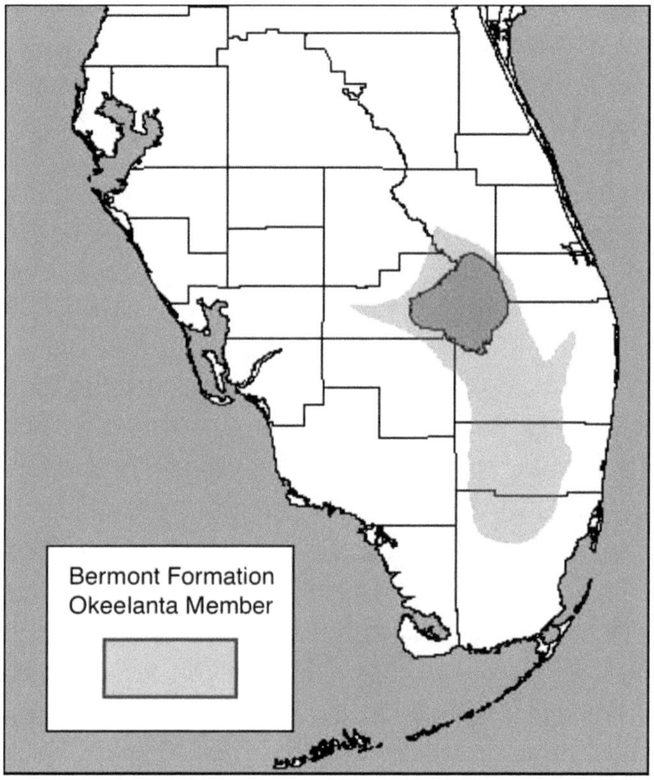

Figure 11.2 Areal distribution of the Okeelanta Member of the Bermont Formation in southern Florida, showing the outline of Pleistocene Lake Okeelanta.

white. In these units, the shell bioclasts are usually stained tan, brown, light gray, or dark bluish-gray. The sandy lime muds are typically of a dark brown or brown-gray color and their molluscan bioclasts are stained a dark brown, dark gray, or black color. The thin limestone units, usually occurring as duricrusts on the bedding plane surface, are colored light brown or light gray and contain white molluscan bioclasts. The Okeelanta Member has an areal extent that conforms to the outline of Lake Okeelanta and has the smallest distribution of the three Bermont members. Outside the boundaries of the paleolake, an unconformity or terrestrial sequence takes the place of the member. Okeelanta sediments are known only from Glades, southern Okeechobee, westernmost Martin, Palm Beach, northeasternmost Hendry, Broward, and northern Dade Counties. In the stratotype area at Okeelanta and in the Holey Land, the member averages 1.5–2 m in thickness (deposited in the deepest part of the lake). In the Palm Beach Aggregates quarries (Figures 10.3 and 10.4), the member was deposited in the peripheral shallow marshland areas and is only 0.5 m thick.

Stratotype. Petuch and Roberts (2007: 164) designated a section on the eastern side of the North New River Canal, 5.2 km south of the city of South Bay, Palm Beach County, as the stratotype of the Okeelanta Member. The type locality is adjacent to the bridge at the intersection of State Highway 27 and County Road 827, northeast of the Okeelanta Corp. sugar mill property. The member crops out in the North New River Canal at a depth of 2.5 m below water surface. Specimens of Okeelanta fossils can be collected on the dirt access road on the top of the canal berm (dredged from the canal and used as road fill). The member takes its name from the present-day Okeelanta sugar mill (Florida Crystals, Inc.) and the old agricultural village of Okeelanta, which was destroyed in the 1928 hurricane.

Age and Correlative Units. Based on the stratigraphic position between the two uranium isotope-dated Bermont members, the Okeelanta Member is now known to date from the early Chibanian Age (previously known as the "Ionian Age") of the middle Pleistocene.

Okeelanta Index Fossils. Present research indicates that the molluscan fauna of the Okeelanta Member is almost completely endemic, with only a few species known to occur outside the paleolake boundaries and with most species being restricted to the lake basin. Here, the different species groups were distributed by ecological preferences, and these assemblages can be used to determine the bathymetry of the paleolake. Only a few of the Okeelanta mollusks have been described to date (shown here), and at least 80 unnamed species remain to be described by future workers. Of these, the planorbid genus *Seminolina* exhibited the largest species radiation, with over 30 different species having been collected. The Okeelanta Member also commonly contains alligator teeth, alligator osteoderms, freshwater turtle shell fragments, and Capybara (*Hydrochoerus*) caviamorph rodent bones and teeth (all found primarily in the Marshland Zone sediments).

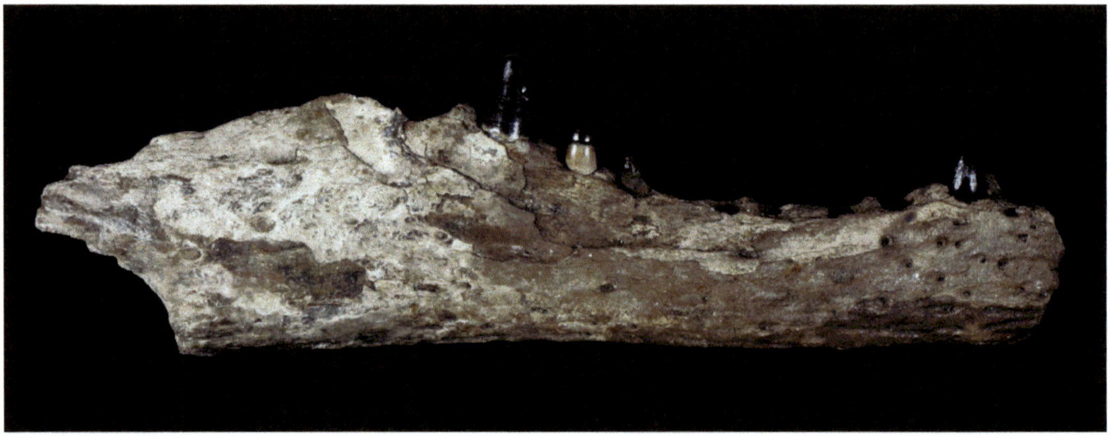

Jaw fragment from an American alligator, *Alligator mississippiensis* Daudin, 1802, from Pleistocene Lake Okeelanta. Length 235 mm. Found in Palm Beach Aggregates Pit No. 6, Loxahatchee, Palm Beach County, Florida.

Figure 11.3 Mollusks of the *Corbicula* Zone of Pleistocene Lake Okeelanta.
A= *Seminolina lindae* Petuch, 1991, length 28 mm; **B**= *Seminolina knepperi* Petuch, 1991, length 15 mm; **C**= *Seminolina roseae* Petuch, 1991, length 16 mm; **D**= *Seminolina wilsoni* (Taylor, 1966), length 22 mm; **E, F**= *Seminolina* cf. *knepperi* Petuch, 1991, length 27 mm; **G**= *Elimia effosa* (M. Smith, 1938), length 18 mm; **H**= *Seminolina* cf. *lindae* Petuch, 1991, length 30 mm; **I**= *Rangia* unnamed species, width 21 mm; **J**= *Seminolina* unnamed species, length 17 mm; **K**= *Corbicula* unnamed species, width 16 mm, specimen embedded (upper left) in Okeelanta freshwater limestone.

Figure 11.4 Fossils of the Marshland Zone of Pleistocene Lake Okeelanta.
A= *Physa barberi* Clench, 1925, length 23 mm; **B**= *Alligator mississippiensis* Daudin, 1802 tooth, length 37 mm; **C**= *Hydrochoerus holmesi* Simpson, 1928 Capybara incisor, length 58.1 mm; **D**= *Apalone* species Soft-Shelled Turtle carapace fragment, length 155.0 mm; **E**= *Stenophysa meigsi* (Dall, 1890), length 28 mm; **F**= *Planorbella preglabrata* (Marshall, 1926), width 28 mm; **G**= *Pomacea* cf. *paludosa* (Say, 1829), length 43 mm; **H**= *Pomacea* cf. *innexa* (Crosse and Bernardi, 1890), length 38 mm; **I**= *Viviparus* unnamed species 1, length 24 mm; **J**= *Viviparus* unnamed species 2, length 22 mm; **K**= *Viviparus* unnamed species 3, length 20 mm.

Belle Glade Member, Bermont Formation. Around 750,000 years B.P., during the Chibanian Age, worldwide climates began to warm and sea level began to rise. With the initiation of this new period of global warming, southern Florida was engulfed by a new marine system, the **Belle Glade Subsea** (named for the city of Belle Glade, Palm Beach County) (Figure 11.5). The rich endemic faunas of early Chibanian Lake Okeelanta quickly perished in the influx of seawater, with only a few species finding refuges in marshlands on higher elevation areas (the ancestors of the Recent Everglades freshwater mollusks). During the preceding cooler time of the latest Calabrian, a large percentage of the Loxahatchee Subsea molluscan fauna became extinct, leaving a much-diminished fauna to repopulate the ecosystems of the Belle Glade Subsea. Some of the more important groups that disappeared from the Florida fossil record during latest Calabrian time included the gastropod genera *Pyruella*, *Malea*, *Pseudozonaria*, and *Jenneria*, and the bivalve genera *Carolinapecten*, *Conradostrea*, and *Armamiltha*. The disappearance of these organisms, and the addition of newly evolved endemic species, gives the Belle Glade fauna a distinctive impoverished appearance that can be used as a stratigraphic marker.

The geomorphology of the Belle Glade Subsea was essentially an extension of the predecessor Loxahatchee Subsea, with the older features being enlarged and better developed during Chibanian time. Immokalee Island had now fused with mangrove forests to the western side of Miccosukee Island, obliterating the shallow lagoon and Turtle Grass environments of the Hendry Platform. Cape Sable Island had now fused to the southern section of the Miami Archipelago, forming a continuous chain of large, wide, pine-forested islands. Likewise, the Palm Beach Archipelago had widened by this time, effectively closing off the Atlantic from the Belle Glade Subsea. Only a few narrow cuts existed between the islands, and these remain today as the Transverse Glades low areas on the Atlantic Coastal Ridge of Palm Beach, Broward, and Dade Counties. Because of the deposition of Lake Okeelanta and its surrounding marshlands during early Chibanian time, the Loxahatchee Trough of the Belle Glade Subsea was much shallower than it was in the older Loxahatchee Subsea. During Chibanian time, only a small, narrow central area retained deepwater conditions, whereas the rest of the trough was composed of broad, shallow carbonate banks (the center of deposition for the Belle Glade Member). To the south, the Monroe Reef Tract was also becoming better developed and extended to at least present-day Bahia Honda Key in the Florida Keys.

Along the northern end of the Belle Glade Subsea, the Bermont Lagoon System was also rapidly filling with sediments during Chibanian time, supporting large cypress forests and coastal marshlands. This explains the diminished presence of Belle Glade Member beds at the Leisey Shell Pit in Hillsborough County. Mangrove forests filled the southern region of the old Kissimmee Embayment, while the northern end had now bcome overgrown with hardwood and pine forests. This explains the absence of Bermont marine beds at the Rucks Pit at Fort Drum,

Figure 11.5 Simulated satellite image (altitude: 100 miles) of the Floridian Peninsula during the Chibanian Pleistocene, showing the probable appearance of the **Belle Glade Subsea** of the Okeechobean Sea. At this time, the **Palm Beach Archipelago**, along the northeastern edge of the subsea basin, had widened and fused into larger islands, effectively closing off the Okeechobean Sea and the entire Kissimmee Embayment was filled with dense mangrove forests.

sediments at this time and contained large carbonate shoals and Turtle Grass beds. Only the Caloosahatchee Strait remained deep, being scoured by strong tidal currents that prevented the accumulation of any thick beds of sediments. This is evidenced by the absence, or extreme truncation, of Bermont beds along the present-day Caloosahatchee River. The environments contained in these geomorphologic features produced the lithofacies and biofacies of the final pulse of Bermont deposition, the Belle Glade Member.

Comprising the upper beds of the Bermont Formation, the Belle Glade Member (named for the city of Belle Glade, Palm Beach County) contained the first set of beds recognized as being mid-Pleistocene in age. As mentioned earlier in this chapter, Druid Wilson (*in* Olsson and Petit, 1964) noticed that the molluscan fossils indicated that this unit fell somewhere between the early Pleistocene Caloosahatchee Formation and the late Pleistocene Fort Thompson Formation. The geochronological placement of the "Glades Unit" and "Unit A" by these early workers, using only molluscan biostratigraphy, was amazingly accurate. From 1964 to 1969, the first large-scale exposures of sediments from the Belle Glade Member were brought up by draglines excavating a road fill rock pit in the city of Belle Glade (Hoerle, 1970; McGinty, 1970). Shirley Hoerle collected and identified 434 species of mollusks from this quarry, attesting to the richness of this upper Bermont fauna (her taxonomy, however, is now outdated). In the 1980s, the senior author encountered the same Belle Glade fauna and sediments during the deepening and widening of the North New River Canal, along Highway 27, 3 to 20 km south of South Bay, Palm Beach County (discussed in the previous section on the Okeelanta Member). While working in the Holey Land area, the senior author also encountered Belle Glade beds at the Griffin Brothers Pit and at the Talisman Canal levee excavations. Similar sediments and molluscan faunas were also found at the Palm Beach Aggregates quarries in Loxahatchee (Figure 10.4) and at the Capeletti Brothers quarries in Miami, demonstrating that this member is widely distributed throughout the Everglades area.

In 1978, Muriel Hunter was the first person to apply a semiformal name to the beds encountered within the Belle Glade rock pit and south of Lake Okeechobee (during dredgings of the Miami and North New River Canals). Considering these beds to be correlative with the Okaloacoochee Member, she named them the "Belle Glade Member" and placed her new member in the late Pleistocene Fort Thompson Formation. Subsequent research has shown that the Belle Glade beds are much older than originally presumed by Hunter and that, cyclostratigraphically and lithologically, they belong in the Bermont Formation. Petuch and Roberts (2007: 168-169) retained Hunter's "Belle Glade Member" but placed it in the chronostratigraphic position as the Chibanian component of Bermont deposition.

Lithologic Description and Areal Distribution. Typically, the Belle Glade Member is composed of thick beds of fine-particulate unconsolidated calcarenites with variable amounts of quartz sand and scattered coral and molluscan bioclasts. Although some

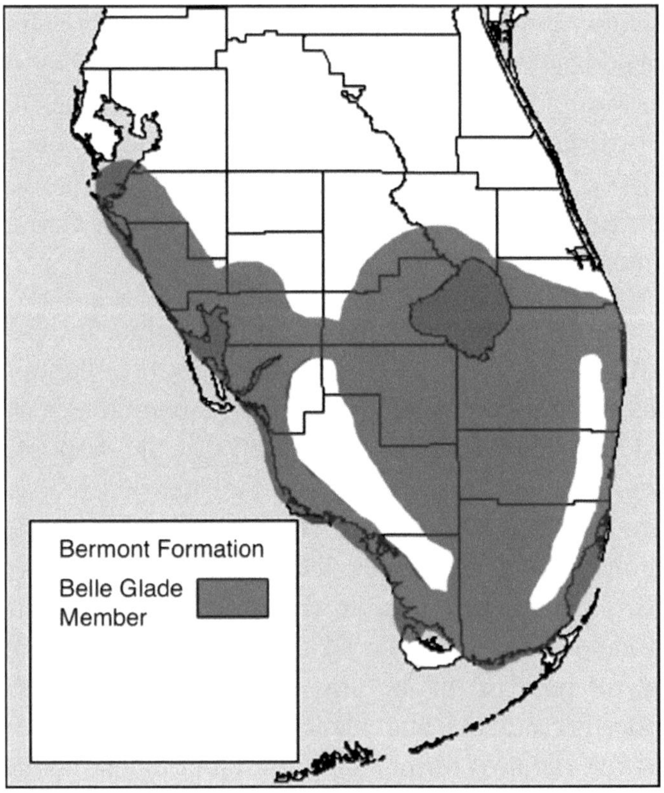

Figure 11.6 Areal distribution of the Belle Glade Member, Bermont Formation, in southern Florida.

beds (such as those seen at Palm Beach Aggregates quarries and along the North New River Canal) contain abundant fossil mollusks, these never form the densely packed masses, with little matrix, seen in the underlying Holey Land Member. Heavily indurated limestones are also absent from this member, although some facies (such as at Palm Beach Aggregates) are partially indurated, forming a friable, weak, sandy limestone (referred to locally as "golden rock"). Some of the calcarenite facies, such as those from the North New River Canal south of South Bay, are very clean, with only a small amount of quartz sand, and contain a coarser-grained calcarenite, almost approaching an oölite in appearance. Some calcarenite beds, such as those exposed along the Talisman Canal on the northern edge of the Holey Land, are extremely fine grained, approaching a calcilutite. The sandy calcarenites vary in color from white, to pale gray, to a light yellow-orange, to dark yellow. Shells preserved in the white to pale gray calcarenites are stained a dark blue-gray or golden-tan color, whereas those in the yellow-orange sandy calcarenites ("golden rock") are white. Equivalent beds outside the Everglades Basin, in cooler water areas such as at the Bermont stratotype at Shell Creek, typically contain mud units and have a much higher percentage of quartz sand and molluscan bioclasts. These sandy shell beds are typically a gray or blue-gray color. Within the stratotype area in western Palm Beach County and at the Capeletti quarries in Miami, large beds of the branching coral *Porites furcata* are frequently encountered in the coarse-grained calcarenite beds. The Belle Glade Member has the same areal distribution as does the older Holey Land Member, extending from southern

Hillsborough County to Monroe County and southern Okeechobee County south to Dade County (Figure 11.6). Throughout this area, the member varies in thickness from 1–3 m at Shell Creek, Charlotte County to 10 m at the Belle Glade rock pit, Palm Beach County, to 3 m in the Palm Beach Aggregates quarries, Palm Beach County.

Stratotype. Retaining Hunter's (1978) original type locality, we here designate the stratotype of the Belle Glade Member to be the old Belle Glade rock pit, at the bend of State Road 80 adjacent to the Glades Memorial Hospital, Belle Glade, Palm Beach County (W ½, NE ¼, Sec. 7, T. 44 S., R. 37 E., Palm Beach County). The quarry is now flooded as a lake, but is still accessible for collecting. The top of the Belle Glade Member crops out within the lake at a depth of 1.8 m below surface and can be accessed by scuba diving.

Age and Correlative Units. Based on the uranium isotope dates discussed earlier in this chapter, the Belle Glade Member is now known to date from the Chibanian Age of the middle Pleistocene. The member is correlative with the upper beds of the Flanners Beach Formation of the Carolinas.

Belle Glade Index Fossils. Although containing a less diverse fauna than did the Holey Land Member, the Belle Glade still has rich and highly endemic molluscan assemblages that can be used to demarcate the member boundaries. Some of the more abundant and indicative Belle Glade index species are shown on the following plates.

Figure 11.7 Close-up of a typical molluscan assemblage from the Belle Glade Member, Bermont Formation, in the DOT fill pit at LaBelle, Hendry County, Florida. Here the strombid *Strombus evergladesensis* can be seen in upper left of the photo, while the muricid *Chicoreus gravesae*, the fasciolariid *Aristofusus watermani*, and the large potamidid *Pyrazisinus roseae* are seen along the bottom of the photo (from left to right). Photo by Michael Bruggeman.

Figure 11.8 Gastropods of the Belle Glade Member, Bermont Formation.
A= *Pusula bermontianus* Petuch, 1994, length 21 mm; **B=** *Strombus lindae* Petuch, 1991, length 44 mm; **C=** *Strombus evergladesensis* Petuch, 1991, length 71 mm; **D=** *Macrostrombus sargenti* Petuch and Drolshagen, 2011, length 220 mm; **E=** *Pyrazisinus miamiensis* Petuch, 1994, length 33 mm; **F=** *Chicoreus gravesae* Petuch, 1994, length 64 mm; **G=** *Vokesinotus griffini* Petuch, 1991, length 37 mm; **H=** *Vokesimurex bellegladeensis* (E. Vokes, 1963), length 47 mm; **I=** *Murexiella graceae* (McGinty, 1940), length 32 mm; **J=** *Cinctura evergladesensis* (Petuch, 1991), length 47 mm.

Figure 11.9 Gastropods of the Belle Glade Member, Bermont Formation.
A= *Polygona jucundus* (McGinty, 1940), length 45 mm; **B=** *Aristofusus watermani* (M. Smith, 1940), length 128 mm; **C=** *Cerithioclava* cf. *garciai* Houbrick, 1985, length 60 mm; **D=** *Chicoreus duerri* Petuch, 1994, length 32 mm; **E=** *Sinistrofulgur roseae* Petuch, 1991, length 238 mm; **F=** *Fulguropsis evergladesensis* Petuch, 1994, length 60 mm; **G=** *Fulguropsis feldmanni* Petuch, 1991, length 72 mm; **H=** *Melongena (Rexmela) diegelae* Petuch, 1994, length 27 mm; **I=** *Miccosukea holeylandica* (Petuch, 1990), length 59 mm; **J=** *Nodipecten pernodosus* (Heilprin, 1886), length 98 mm.

Figure 11.10 Gastropods of the Belle Glade Member, Bermont Formation.
A= *Lindaconus lemoni* (Petuch, 1990), length 57 mm; **B=** *Calusaconus evergladesensis* (Petuch, 1991), length 29 mm; **C=** *Perplexiconus palmbeachensis* (Petuch, 1994), length 25 mm; **D=** *Americoliva murielae* (Olsson, 1967), length 41 mm; **E=** *Americoliva lindae* (Petuch, 1991), length 28 mm; **F=** *Americoliva edwardsae* (Olsson, 1967), length 30 mm; **G=** *Americoliva cokyae* (Petuch, 1991), length 16 mm; **H=** *Lindoliva diegelae* Petuch, 1988, length 51 mm; **I=** *Turbinella hoerlei* E. Vokes, 1966, length 270 mm; **J=** *Americoliva southbayensis* Petuch, 1994, length 40 mm.

CHAPTER 12. The Lake Worth Subsea (Fort Thompson Formation)

After a 400,000-year period of warm, subtropical conditions within the Belle Glade Subsea, sea levels dropped precipitously, and the entire Florida Platform was again emergent. Caused by the buildup of immense continental glaciers around the Northern Hemisphere, this sea level drop marks the beginning of both the Tarantian Age and the late Pleistocene. Like the previous Chibanian Age, the Tarantian was complex, with at least two glacial pulses separated by a short, warmer interglacial time. The first glaciation, at around 126,000 years BP, was devastating and produced the lowest sea level seen since the Oligocene. Much of the rich and endemic Belle Glade marine fauna disappeared at this time and the resultant marine fauna was highly impoverished. As a result of the large amount of marine sediment deposition during Chibanian time, the deeper areas within the Loxahatchee Trough were completely infilled by Tarantian time, producing a broader, wider, and shallower basin than existed previously. During Tarantian low sea levels, the Loxahatchee Trough area was covered with a system of interconnected ephemeral marshlands and small lakes, establishing the framework for the modern Everglades.

By the middle of Tarantian time, a major sea level rise occurred, flooding the Okeechobean Sea basin with very shallow marine and estuarine environments. This sea level rise marked the beginning of the **Lake Worth Subsea** (named for the city of Lake Worth, Palm Beach County; see Petuch, 2004), the last depositional subdivision of the Okeechobean Sea (Figure 12.1). Within 100,000 years of being inundated with marine environments, the sea withdrew and southern Florida was again exposed to subaerial, terrestrial conditions. This occurred during the second, and more severe, glacial pulse at the end of Tarantian time. Throughout this regressional episode, Florida experienced dry, semiarid conditions and the Everglades area more closely resembled east central Texas than it does the present-day marshlands. A large, shallow freshwater body, Lake Pahayokee (see Petuch, 2004: 247), formed in the northern part of the Everglades Basin but lasted for only a short period of a few thousand years. This last great paleolake was populated with the descendants of the survivors from Lake Okeelanta. At the beginning of the Tarantian Age, and after the lowest sea level drop of the Neogene period, the worldwide oceanic system quickly rebounded and flooded the continental margins with the highest known sea level, possibly 20 m above that of the present. This eustatic high, at 120,000 years BP, ushered in the greatest, development of the Lake Worth Subsea. Together, the early and late Tarantian sea level rises produced the environments for the deposition of the last extensive Everglades geologic unit, the **Fort Thompson Formation** (with the early Tarantian Okaloacoochee Member and the late Tarantian Coffee Mill Hammock Member).

The higher sea level stand of the late Tarantian flooded the Kissimmee Embayment to the greatest extent since the Gelasian, allowing mangrove trees to grow as far north as present-day northern Okeechobee County. A reddish, peaty facies of the Coffee

Figure 12.1 Simulated satellite image (altitude: 100 miles) of the Floridian Peninsula during the Tarantian Pleistocene, showing the probable appearance of the **Lake Worth Subsea** of the Okeechobean Sea. At this time, the entire subsea basin had filled with sediments, allowing dense mangrove forests to grow along the western half and creating the very shallow, estuarine **Seminole Lagoon System** along the eastern half. A new and larger coral reef system formed on top of the older Bermont-aged Monroe Reef Tract, the **Key Largo Reef System**, and this still exists today as the northern half of the Florida Keys. Large oölite deposits, the **Miami Oölite Banks**, formed behind the Key Largo Reef System and produced the oölitic limestones of Miami and the southern Florida Keys.

Mill Hammock Member of the Fort Thompson Formation was deposited within these mangrove environments and these conspicuous fossiliferous sediments, often containing carbonized mangrove wood, were prominent within the Rucks Pit at Fort Drum. Along late Tarantian western Florida, a narrow set of coastal lagoons formed behind barrier islands and marked the incipiency of the Recent Gulf Lowlands and Gulf Barrier Islands. The thick and well-developed shell beds of the Fort Thompsonnear Venice, Sarasota County, were deposited in the extensive Turtle Grass beds that grew within these lagoon systems.

Fort Thompson Formation, Okeechobee Group. In his publication on stratigraphic sections across the Everglades, Sellards (1919) named a distinct set of mixed lithologies seen along the Caloosahatchee River near LaBelle, Hendry County, the "Fort Thompson beds." These beds, composed of freshwater, estuarine, and shallow marine deposits, were overlaid by a thick layer of "shell marl" composed almost entirely of *Chione* clams. This distinctive unit was named the "Coffee Mill Hammock marl" in the same publication. In their review of Florida geology 10 years later, Cooke and Mossom (1929) elevated the "Fort Thompson beds" to formational status and considered the Coffee Mill Hammock to be the upper member of their new formation. In this book, we concur with Cooke and Mossom and keep the "Fort Thompson beds" and the "Coffee Mill Hammock marl" together as a single formation. With the exception of Brooks (1968) and Lyons (1991), who consider them separate formations, most workers in the Everglades area follow Cooke and Mossom and consider them to represent a single formation. In his study of the stratigraphy of the Caloosahatchee River area, DuBar (1958) observed that the lower Fort Thompson beds, below the Coffee Mill Hammock Member, were more lithologically complex than described by earlier workers. These contained a mixture of fresh and brackish water deposits and a distinctive layer of scallops, the "*Chlamys* Bed" (now placed in the genus *Argopecten*). DuBar (1958a) later named this lower section of the Fort Thompson Formation the Okaloacoochee Member, after the Okaloacoochee Branch that empties into the Caloosahatchee River near LaBelle.

Lithologic Description and Areal Distribution. The Fort Thompson typically consists of thin, intercalated beds with three main lithologies: freshwater calcarenites ("marls") with *Seminolina* planorbid gastropod bioclasts, densely packed beds of the bivalve *Chione elevata* (previously referred to as *Chione cancellata*, which comprises the majority of the carbonates) in a sparse quartz sand matrix, and muddy sand with beds of bivalves, either the mactrid *Rangia cuneata* (Figure 7.3) or the scallop *Argopecten irradians* subspecies. The scallops are typically stained dark gray or black, whereas the *Rangia* and *Chione* are white or pale tannish-white. Locally, the freshwater beds are completely or partially indurated, forming thin limestone units. In the south, under the Atlantic Coastal Ridge in Dade County (subjacent to the Miami Formation), the lower member of the Fort Thompson is present and is composed of a yellow or pale tan sandy limestone filled with the molds of dissolved mollusk shells.

Some of these molds are filled with calcite, producing perfect calcitic pseudomorphs of gastropods and bivalves. This same yellow sandy moldic limestone also frequently contains heavily recrystallized corals, mostly the rose coral *Manicina* and the star coral *Montastrea*. The quartz sands of the *Chione* beds are white, yellow, or pale tan, whereas those mixed with the fine calcarenites of the *Rangia* and *Argopecten* beds are light or dark gray. At Palm Beach Aggregates quarries at Loxahatchee, the Fort Thompson is indurated into a cream-yellow sandy limestone with abundant molluscan bioclasts, mostly *Chione*. In the northern part of the Okeechobee Plain, within the Kissimmee River Valley (particularly the Fort Drum area), the Fort Thompson is composed of a reddish-orange sandy peat filled with the gastropod *Melongena corona* and the bivalve *Pseudomiltha floridana*. This lithology represents deposition within the organic-rich estuarine sand flats of the Kissimmee Embayment mangrove forests.

The combined ranges of the two members of the Fort Thompson Formation cover the entire Everglades area, from Pinellas and Hillsborough Counties to Monroe County and from northern Highlands and Okeechobee counties to Dade County. The Okaloacoochee Member, deposited during a relatively low sea level stand, is less extensive than the Coffee Mill Hammock Member, which was deposited during the Sangamonian extreme sea level high. The formation, in total, varies in thickness from over 3 m at the Leisey Shell Pit in Hillsborough County (Morgan and Hulbert, 1995) to 5 m at the Venice Shell Pit in Sarasota County, to 2.5 m at the type section in Hendry County, to 1 m along the Miami Canal and at Loxahatchee, Palm Beach County, to 5.5 m at the Six-Mile Bend on old State Road 80 east of Belle Glade, Palm Beach County, to 1.5 m at the Capeletti Brothers Pits west of Miami, Dade County. At most localities, the upper surface of the Coffee Mill Hammock lies between 0.5 and 2.5 m below land surface.

Stratotype. From Cooke (1945: 254): "Old Fort Thompson, the type locality of the formation, lay on the south bank of the Caloosahatchee, 1½ mile east of the courthouse at LaBelle." This is essentially the same locality as DuBar's station A15 (DuBar, 1958: 231): "SE ¼, SW ¼ Sec. 3, T. 43 S., R. 29 E., Hendry County, Florida, left bank of the Caloosahatchee River approximately 1.5 miles east of the bridge at LaBelle." The top of the formation lies 1 m below land surface.

Age and Correlative Units. Fort Thompson deposition extends from mid-Tarantian Age (approximately 120,000 years B.P.) to the late Tarantian Age (75,000 years B.P.). The formation is correlative with the Canepatch and Socastee Formations of the Carolinas.

Fort Thompson Index Fossils. Reflecting the time differential between the two members of the Fort Thompson Formation, the compositions of the marine molluscan faunas of the Okaloacoochee and Coffee Mill Hammock differ greatly. Because of the paucity of large exposures of the Okaloacoochee, its fauna was thought to be essentially the same as the Coffee Mill Hammock by most previous workers. Subsequent research in extensive new Okaloacoochee exposures, such as at Parkland,

Broward County (Petuch, 1994, 2004; Petuch and Roberts, 2007), has shown that the two members have recognizably distinct faunas. The index fossils are listed in the respective section for each member.

Okaloacoochee Member, Fort Thompson Formation. As discussed previously, the lower section of the Fort Thompson Formation was first noticed and described from exposures along the Caloosahatchee River in Hendry County (after the river was channelized by the Army Corps of Engineers). DuBar (1958a, 1974) formally named these lower beds, including his "*Chlamys* Bed," the Okaloacoochee Member for the Okaloacoochee Branch, a small tributary of the Caloosahatchee in Hendry County. He also dated the member as being "early Sangamonian" in age (now known as the early Tarantian Age) and the overlying Coffee Mill Hammock being "late Sangamonian" in age (now known as the late Tarantian Age).

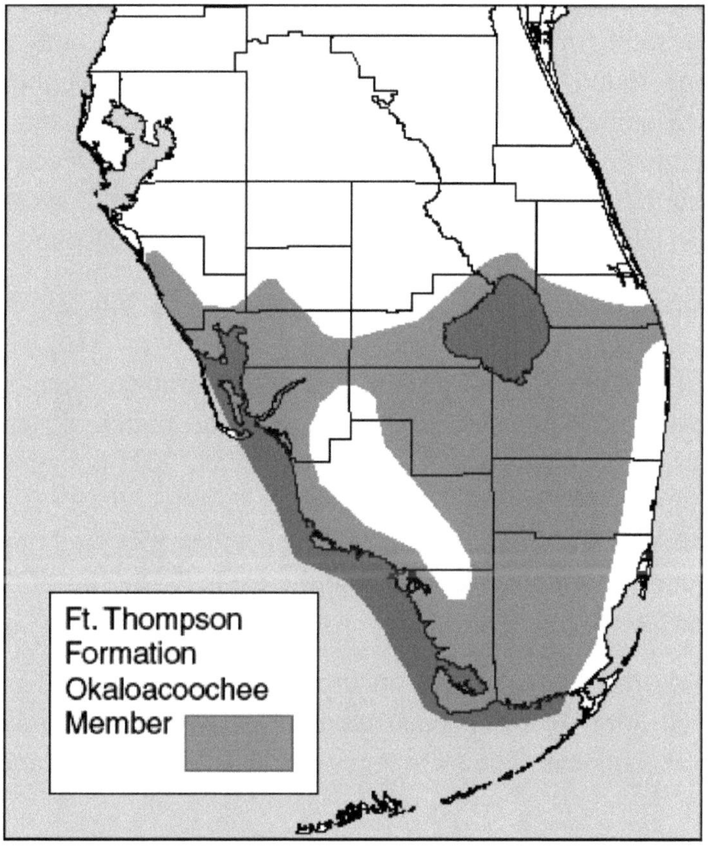

Figure 12.2 Areal distribution of the Okaloacoochee Member, Fort Thompson Formation, in southern Florida.

Lithologic Description and Areal Extent. Condensed here is an extract from DuBar (1974: 224): "The Okaloacoochee Member is composed of freshwater marl locally divided into upper and lower units by a thin tongue of brackish water marl … The freshwater beds are slightly to moderately indurated; however, locally, the upper few inches to one foot are penetrated by vertical solution holes which commonly are filled with the matrix of overlying units … The clastic fraction of the marls ranges up to 33% by weight, and consists primarily of fine to medium quartz sand. The marl is generally white to light gray or cream in color, but locally it may be yellowish brown or tan … Fossils are abundant and well preserved … In the type area and along Banana Creek, a tongue of brackish water shelly sand four to six inches thick separates the freshwater marl into upper and lower units … The most characteristic species of this unit is the bay scallop *Chlamys irradians concentricus* (now known as *Argopecten irradians* unnamed subspecies). The clastic fraction, which is less than 50% by weight, consists of subangular to subrounded medium quartz sand and minor amounts of silt and clay." In the northern Everglades, particularly the region south of Lake Okeechobee and the Everglades Agricultural Area (EAA), thin beds of the brackish water bivalve *Rangia cuneata* are found intercalated between freshwater calcarenite layers, indicating rapidly fluctuating salinities within the **Seminole Lagoon System** during Okaloacoochee time. The member extends from Sarasota County to Monroe County and from southernmost Okeechobee County to Dade County (the yellow sandy moldic limestone found below the Miami Formation). Throughout its range, the Okaloacoochee Member varies in thickness from 2 m at the stratotype in Hendry County to 3 m at the Six-Mile Bend east of Belle Glade, Palm Beach County, to 4 m at Parkland, Broward County, to 1 m along the Miami Canal, western Palm Beach County, to 2 m in the County Pit at Okeechobee, Okeechobee County.

Stratotype. From Dubar (1958: 224): "The member is typically developed on the Caloosahatchee River near Okaloacoochee Branch in Hendry County." The Okaloacoochee Branch is a small stream, tied into a drainage canal system that is fed by flowage from the Okaloacoochee Slough and the marsh system of Collier and Hendry Counties. DuBar's station A14 (DuBar, 1958: 231) contains the best section from the stratotype area: "SW ¼, SW ¼ Sec. 2, T. 43 S., R. 29 E., Hendry County, Florida, right bank of the Caloosahatchee River about 150 yards upstream from the point of intersection of a large meander and the canal." The top of the member is 1.5 m below land surface."

Age and Correlative Units. Based on uranium isotope dating, as summarized by Lyons (1991: 160), the Okaloacoochee Member is now known to date from the mid-to-late Illinoian Pleistocene. The member is correlative with the Canepatch Formation of the Carolinas.

Okaloacoochee Index Fossils. The Okaloacoochee contains a molluscan fauna that is transitional between that of the Belle Glade Member of the Bermont Formation and the Coffee Mill Hammock Member of the Fort Thompson Formation. Although it has a high percentage of taxa that are still extant, particularly the bivalves, this member contains a number of endemic taxa and relicts from Bermont time. Of particular interest is the presence of the gastropod genus *Turbinella* and a surviving descendant of the dwarf strombids of the Bermont Formation (the *Strombus lindae* species complex). Both these taxa became extinct in Florida before Sangamonian times and are absent from the Coffee Mill Hammock Member. The marine communities of the Okaloacoochee Member were described by the senior author (Petuch, 2004: 249–251) and include the ***Pyrazisinus gravesae*** **Community** (mangrove forests and intertidal mud flats) and the ***Strombus alatus kendrewi*** **Community** (shallow lagoon sand flats and Turtle Grass beds). Some of the more abundant and indicative Okaloacoochee index fossils are shown here in the following figure.

Cabbage Palmetto silicified ("petrified") wood (*Sabal palmetto*), length 252 mm. Silicified wood from palmetto trees is common in some facies of the Okaloacoochee Member.

Figure 12.3 Gastropods of the Okaloacoochee Member, Forth Thompson Formation.
A= *Strombus alatus kendrewi* Petuch, 2004, length 74 mm; **B=** *Strombus lindae dowlingorum* Petuch, 2004, length 51 mm; **C=** *Macrostrombus costatus griffini* (Petuch, 1994), length 98 mm; **D=** *Sinistrofulgur perversum okeechobeensis* Petuch, 1994, length 114 mm; **E=** *Americoliva sayana sarasotaensis* Petuch and Sargent, 1986, length 41 mm (extant in deeper water areas off western Florida); **F=** *Jaspidiconus pfluegeri* Petuch, 2004, length 22 mm; **G=** *Fulguropsis pyruloides pahayokee* Petuch, 1994, length 88 mm; **H=** *Turbinella wheeleri* Petuch, 1994, length 185 mm (extant along the coast of Yucatan, Mexico); **I, J=** *Macrocypraea (Lorenzicypraea) cervus* (Linnaeus, 1771), length 115 mm.

Coffee Mill Hammock Member, Fort Thompson Formation. This uppermost member of the Fort Thompson Formation was described from outcrops near the Ortona Lock of the Caloosahatchee River, Glades County, by Sellards (1919). Although he considered it to be a separate formation, most subsequent workers have included it within the Fort Thompson. North of the St. Lucie Inlet of Martin and St. Lucie Counties, the typical Coffee Mill Hammock carbonate bioclast lithology disappears and is replaced by a dark gray sand containing cooler water mollusks and no *Chione* clams. The molluscan fauna of this unit, which may represent a southern extension of the Socastee Formation of South Carolina, is composed of the gastropods *Busycon carica*, *Busycotypus canaliculatum*, *Neverita duplicata*, and *Americoliva sayana* and the bivalves *Mercenaria mercenaria* and *Crassostrea virginica*. This assemblage is still living along the coasts of the Carolinas, Georgia, and northern Florida. The presence of *Busycon carica* along the coast of St. Lucie and Martin Counties during Tarantian time demonstrates that the open oceanic water temperature was slightly cooler than it is today. As during Gelasian time, only the Okeechobean Sea basin remained tropical during the late Tarantian.

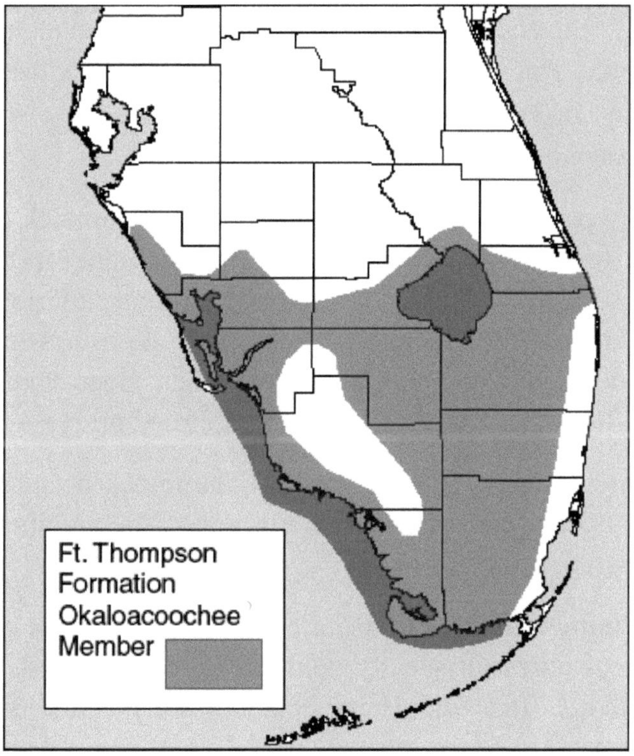

Figure 12.4 Areal distribution of the Coffee Mill Hammock Member, Fort Thompson Formation, in southern Florida.

Lithologic Description and Areal Distribution. Cooke (1945: 249) describes it as "a thin shell marl composed chiefly of *Chione cancellata*" [authors' note: this species is now referred to as *Chione elevata*, with the true *C. cancellata* being a related but different Caribbean species]. DuBar (1974: 223) gives a more detailed description:

"The fossiliferous facies of the member contains abundant and well-preserved molluscan assemblages in a matrix of fine, medium-tan to gray quartz sand. In the type area, the clastic fraction comprises about 80% of the unit, and the carbonate fraction is primarily in the form of fossil shells. The most common and typical species of the Coffee Mill Hammock are the pelecypod *Chione cancellata* [now *Chione elevata*] and *Transenella conradiana* and the gastropods *Bulla occidentalis* and *Olivella mutica*." We here expand the typical *Chione* bed description of the stratotype to include other lithologies such as a reddish-orange peaty sand (mangrove peat) with abundant molluscan bioclasts, usually the bivalve *Pseudomiltha floridana* and the gastropod *Melongena (Rexmela) corona*; a white quartz sand with abundant molluscan bioclasts, usually *Cerithium muscarum*, *Pyrazisinus ultimus*, and *Cerithidea* species, and occasionally present as a pale yellow sandy moldic limestone with partially dissolved mollusks, mostly *Chione elevata*. The reddish peaty facies was deposited within the late Tarantian Kissimmee Embayment and is the dominant facies along the northern Okeechobee Plain. The Coffee Mill Hammock Member varies in thickness from over 3 m at the Leisey Shell Pit, Hillsborough County, to 3 m in the County Pit, Okeechobee, Okeechobee County, to 1 m in the Palm Beach Aggregates quarries, Loxahatchee, Palm Beach County, to 3 m along State Road 7 (U.S. 441) in Lake Worth, Palm Beach County, to 2.5 m in the Rucks Pit, Fort Drum, Okeechobee County. In Monroe and Dade Counties, the Coffee Mill Hammock grades into the oölitic limestones of the Miami Formation.

Stratotype. The type locality of the Coffee Mill Hammock Member is on the Caloosahatchee River near the Ortona Lock, Glades County. (Dubar's station: A57, NE ¼, SW ¼ Sec. 27, T. 42 S., R. 30 E., Glades County, Florida.) "This is the type locality of the Coffee Mill Hammock marl. The exposures from which a large collection was made occur along the right bank of the Caloosahatchee River about 300 yards downstream from the Atlantic Coast Line Railroad bridge" (Dubar, 1958: 240).

Age and Correlative Units. The Coffee Mill Hammock Member is now known to date from the late Tarantian Age of the Pleistocene. The member is correlative with the Socastee Formation of the Carolinas.

Coffee Mill Hammock Index Fossils. This is the uppermost geologic unit of the Everglades area that characteristically contains densely-packed, highly-fossiliferous beds. The distinctive Coffee Mill Hammock molluscan fauna is composed of three separate components: **1.** species that are still living in southern Florida; **2.** living Caribbean species now extinct in Florida; and **3.** Coffee Mill Hammock endemic species that are now extinct. The marine communities of the Coffee Mill Hammock Member were described by the senior author (Petuch, 2004: 250–253) and include the ***Pyrazisinus ultimus*** **Community** (mangrove forests and intertidal mud flats) and the ***Chione elevata*** **Community** (shallow lagoon and Turtle Grass beds). Some of the more abundant and indicative species are shown on the following figures.

Figure 12.5 Index Fossils of the Coffee Mill Hammock Member, Fort Thompson Formation.
A, B= *Cassis spinella* Clench, 1944, length 185 mm; **C=** *Triplofusus papillosus* (Sowerby I, 1825), length 282 mm; **D=** *Aliger (Eustrombus) gigas* (Linnaeus, 1758), length 178 mm; **E=** *Americoliva sayana* (Ravenel, 1834), length 72 mm; **F=** *Melongena (Rexmela) corona winnerae* Petuch, 1994, length 150 mm; **G=** *Vasum muricatum* (Born, 1778), length 96 mm; **H=** *Gradiconus burryae* (Clench, 1942), length 49 mm.

Figure 12.6 Gastropods of the Coffee Mill Hammock Member, Fort Thompson Formation. **A**= *Strombus alatus* Gmelin, 1791, length 82 mm; **B**= *Melongena melongena* (Linnaeus, 1758), length 109 mm (now living in the Caribbean and southern Gulf of Mexico; extinct in Florida); **C**= *Pyrazisinus ultimus* Petuch, 2004, length 53 mm (last species of *Pyrazisinus*; genus now extinct); **D**= *Melongena (Rexmela) corona* (Gmelin, 1791), length 82 mm; **E**= *Melongena (Rexmela) bispinosa* (Philippi, 1844), length 41 mm (extant along the Yucatan Peninsula of Mexico, where it is endemic); **F**= *Lobatus raninus* (Gmelin, 1791), length 58 mm; **G**= *Lindaconus atlanticus* (Clench, 1942), length 62 mm; **H**= *Hexaplex fulvescens* (Sowerby II, 1834), length 143 mm.

CHAPTER 13. The Lake Worth Subsea (Key Largo, Miami, and Anastasia Formations)

The Lake Worth Subsea of the Tarantian Age represented the last vestige of the Okeechobean Sea, and the culmination of 40 million years of shallow marine deposition in the Everglades area. During this time, three main marine environments dominated the shallow areas of the Lake Worth Subsea (Figure 12.1 in Chapter 12); **1. Turtle Grass beds** (*Thalassia testudinum*) and their accompanying beds of the small venerid bivalve, *Chione elevata*; **2. Zonated Coral Reef Tracts**; and **3. Oölite Shoals** and oölite submarine dunes. Much of the **Fort Thompson Formation** (discussed previously in Chapter 12), especially the Coffee Mill Hammock Member, was deposited in Turtle Grass environments and these sea grasses and *Chione* beds covered much of the Lake Worth Subsea sea floor. At the same time that the sediments from these sea grass and shell beds were being deposited, a large and extensive coral reef system was developing farther south, along Dade and Monroe Counties. This reef tract (discussed later in this chapter) produced the massive reefal limestones of the **Key Largo Formation** and these form the underpinnings of the modern northern and central Florida Keys (both High and Low Coral Keys). The high-energy surf conditions along these reef tracts created the perfect environment for the production of calcareous oöids, forming in the surf zone of the reef crest of the Key Largo Reefs. These small pelletoids then later washed back into the lagoon areas to the west and fanned out to create the wide oölitic platform of the **Miami Formation** (discussed later in this chapter). These dune-like deposits, similar to those seen on the present-day Bahama Banks, extended across Florida Bay and into the Gulf of Mexico, forming the oölite islands of the lower Florida Keys. All three of these contemporaneous formations actually represent different depositional environments of the same paleosea, essentially facies of a single large sedimentary unit. In keeping with traditional Florida geology, however, we here retain their individual formational status. The Fort Thompson, Key Largo, Miami, and Anastasia Formations represent the final deposition of the carbonate-rich sediments of the Okeechobee Group

Key Largo Formation, Okeechobee Group. In his survey of the topography and surficial geology of southern Florida, Sanford (1909) first described the coral limestone found in the near surface of the northern and middle Florida Keys. This was named the "Key Largo limestone" for Key Largo, the longest of the Florida Keys. As the geologist assigned to the construction of the Florida East Coast Railway, Sanford had unprecedented opportunities to study sections of the Key Largo Formation that were exposed in canal walls and manmade cuts between keys. In their classic study of the geology and origins of the Florida Keys, Hoffmeister and Multer (1968) were the first to describe the detailed lithologic and paleoecological features of the Key Largo as well as its thickness and structure. The older Ionian-aged Monroe Reef Tract had enlarged and developed into the **Key Largo Reef System** by Tarantian Pleistocene time (named for Key Largo, Monroe County; see Petuch and Roberts, 2007), and this became the predecessor of the present-day Florida Keys. The high-energy zonated reefs of the Key Largo Reef System produced the massive coral limestones of the Key Largo Formation.

Lithologic Description and Areal Distribution. The following is a condensed extract from the lithologic summary of DuBar (1974: 230): "The limestone is composed of an in situ organic framework of hermatypic corals and trapped interstitial calcarenite … cementing and binding organisms, such as crustose coralline algae and milleporid corals, form only a small portion of the reef … the coral heads are surrounded by poorly sorted calcarenite and that well-sorted calcarenite predominates in channels … aragonite and high magnesium calcite probably were the chief original mineralogic components. The original high-magnesium fossils have been altered in place to low-magnesium calcite through the removal of excess magnesium within the meteoric zone. Aragonite is being dissolved currently by meteoric water, and the dissolved calcium carbonate is being redeposited near the site of dissolution as low-magnesium calcite cement. Many originally aragonitic fossils now occur as molds of sparry calcite … five hermatypic corals account for 30% of the rock volume … *Montastrea annularis* is by far the most important coral species. The coral limestone varies in color from white, to pale yellow, to light gray."

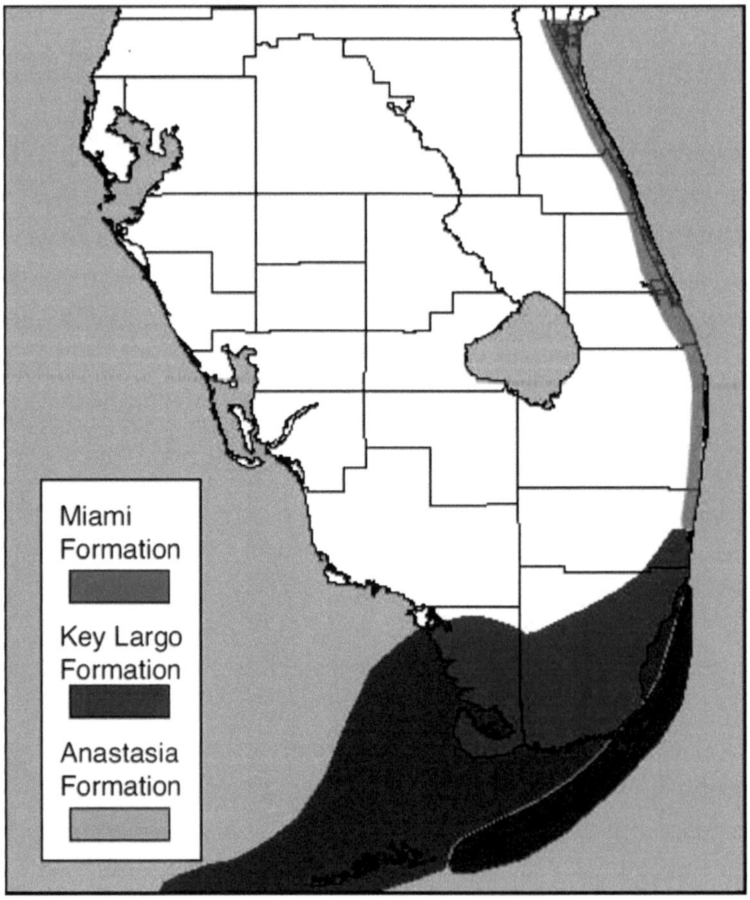

Figure 13.1 Areal distribution of the Key Largo, Miami, and Anastasia Formations in southern Florida.

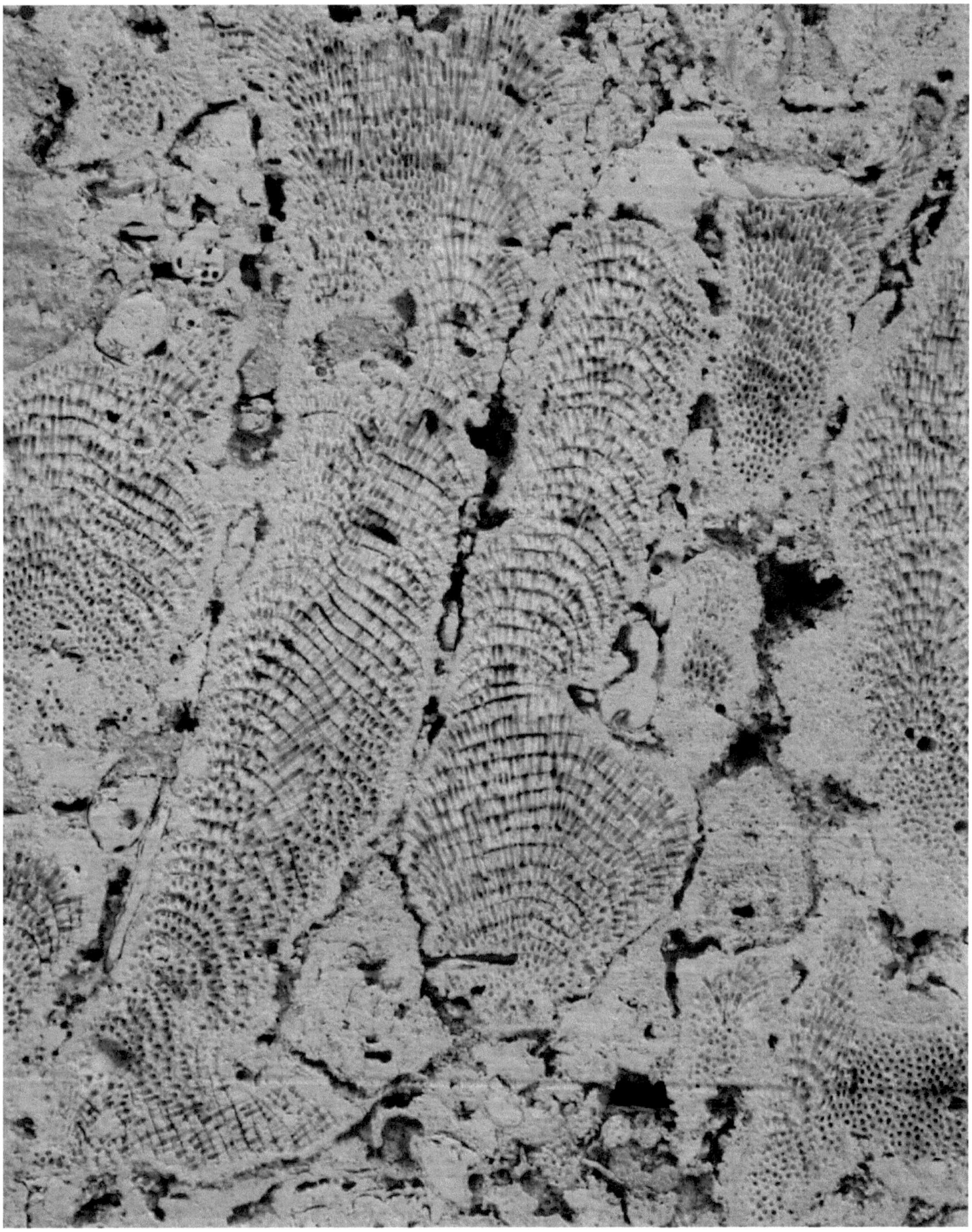

Figure 13.2 Detail of an exposure of the Key Largo Formation in the abandoned quarry on Windley Key, Upper Florida Keys, Monroe County, Florida, showing cross sections of several heads of the massive Star Coral *Montastrea annularis*. This quarry is considered to be the stratotype for the Key Largo Formation. Photo by Anton Oleinik.

As shown in Figure 13.1, the Key Largo Formation underlies only the High and Low Coral Keys of the southeasternmost Dade and Monroe Counties, from Key Biscayne and Soldier Key to Bahia Honda Key. Throughout this range, the fossil reef tract is exposed above water in a narrow strip of 5 km width. The Key Largo extends at least another 8 km seaward, to the edge of the modern living reef tract. On the Florida Bay side, the Key Largo interfingers with the oölites of the Miami Formation. The formation ranges in thickness from 20–50 m, with the thickest limestones being in the western part. The older, lower calcarenite and calcilutite facies, occurring 60–70 m below Key Largo and Key West (Hoffmeister and Multer, 1968), were part of the Monroe Reef Tract and date from the mid-Pleistocene. These represent a reefal facies of the Bermont Formation (see Chapters 10 and 11).

Stratotype. The type section for the Key Largo Formation is in the Windley Key Quarry, now a geologic preserve, on Windley Key, between Plantation Key and Upper Matecumbe Key (Figure 13.2).

Age and Correlative Units. Based on uranium isotope dating (Osmond et al., 1965), the upper part of the Key Largo Formation is now known to have been deposited 120,000 years B.P., placing it in the early Tarantian Pleistocene. This upper section is correlative with the Miami Formation, the Anastasia Formation, and the Coffee Mill Hammock Member of the Fort Thompson Formation, all of southern Florida, and the Socastee Formation of the Carolinas. The lower section of the Key Largo Formation dates from the earliest Tarantian Pleistocene and is correlative with the Okaloacoochee Member of the Fort Thompson Formation of southern Florida and the Canepatch Formation of the Carolinas.

Key Largo Index Fossils. Only five species of scleractinian corals, all massive forms, dominate the Key Largo Formation. These include the following: *Diploria clivosa*, *Diploria labyrinthiformis*, *Diploria strigosa*, *Montastrea annularis*, and *Porites astreoides*. All of these are reef platform and back-reef species, indicating that the reef crest existed farther seaward, under the present-day Hawk Channel. Hoffmeister and Multer (1968) discovered the reef crest species *Acropora palmata* in cores taken from the outer edge of the platform, demonstrating the existence of this reef zone. The Key Largo reef crest, however, was largely removed by wave action and dissolution during the Wisconsinan sea level lows, leaving the reef platform and back reef as the only remnants of a once-wider feature. At the Windley Key Quarry, the encrusting bivalves *Chama macerophylla* and *Spondylus americanus* can be seen to have grown on the eroded undersides of some of the massive coral heads.

Miami Formation, Okeechobee Group. As discussed previously under the description of the Lake Worth Subsea, the Miami Formation is the result of deposition of oöids produced on, and behind, the Key Largo Reef System during Tarantian time. These accumulated in immense dune-like structures on the Miami Oölite Banks (see

Figure 12.1). The first geologist to study the limestones of the Atlantic Coastal Ridge and Florida Keys, Samuel Sanford (1909), named the exposed sections at Miami the "Miami oölite" and that of the central and southern Florida Keys the "Key West oölite". He separated the two formations by differing amounts of quartz sand, with the Miami having a higher percentage than the Key West. Recognizing that the two limestones were of the same age and that they formed a lithologic continuum, Cooke and Mossom (1929) later combined the Key West with the Miami, considering it to be simply a facies. Studying analogue environments in the Bahamas, Hoffmeister et al. (1967) later recognized two broad lithofacies and biofacies within the Miami Formation: an oölitic facies composed of unconsolidated, partially consolidated (by calcite microspar cementing), or indurated (oömoldic) limestones and a "bryozoan facies" made up primarily of interlocking branches of the cheilostome ectoproct ("bryozoan") *Schizoporella floridana*. The oölitic "dunes" of the Miami area (particularly near Coconut Grove) are frequently cross-bedded (Figure 13.4), with dips of up to 30°.

Lithologic Description and Areal Distribution. The best lithologic description of the Miami Formation, summarizing previous works, was given by DuBar (1974: 228; repeated here in condensed form): "...The fresh rock is relatively soft, but hardens upon exposure to the atmosphere. The chief constituent particles are oöids, pellets, and skeletal sand ... Many of the originally aragonitic oöids and pellets have been replaced by calcite. With depth they are increasingly embedded in a crystalline calcite matrix. Below the water table, both oöids and pellets have completely dissolved away, leaving only spherical and ellipsoidal cavities so that the rock becomes oömoldic. The calcitic matrix is a post-depositional precipitation filling the empty interstices ... In the Atlantic Coastal Ridge, the bryozoan facies underlies the oölitic facies ... the facies consists predominantly of zoaria of the cheilostome bryozoan *Schizoporella floridana* (up to 80% by volume). The bryozoans are mixed with oöids, pellets, skeletal sand, polychaete worm tubes, and a sparse molluscan assemblage ... The hard calcitic matrix forms a cellular or vesicular structure." The Miami Formation extends from extreme southeastern Broward County to most of Monroe County. In Monroe County, the Miami underlies Cape Sable, all of Florida Bay, and the Oölite Keys (Big Pine Key area to west of Key West). Within this range, the bryozoan facies makes up most of the lithofacies, covering an area of over 3000 km2. The Miami Formation averages 2–5 m thickness throughout most of its areal distribution.

Stratotype. The type locality of the Miami Formation is generally considered to be the exposure along Silver Bluff, on Bayshore Drive in Coconut Grove, Miami, Dade County.

Age and Correlative Units. As summarized by DuBar (1974: 229), uranium isotope dating of the Miami Formation shows that it ranges in age from 120,000–130,000 years B.P., placing it in the early Tarantian Age of the late Pleistocene. The formation is correlative with the Socastee Formation of the Carolinas, the Coffee Mill Hammock

Member of the Fort Thompson Formation of southern Florida, the Anastasia Formation of eastern Florida, and the upper beds of the Key Largo Formation of southern Florida.

Miami Index Fossils. Although branching colonies of the cheilostome ectoproct *Schizoporella floridana* make up most fossils preserved in the Miami Formation, rare specimens of gastropod and bivalve mollusks do occur. These are primarily the large queen conch *Aliger (Eustrombus) gigas* and the bivalve *Codakia orbicularis*. Both species are common in Recent Caribbean and southern Floridian Turtle Grass beds. DuBar (1974: 228) comments that "two common growth forms are exhibited by the bryozoan colonies. The forms on the Atlantic Coastal Ridge area are rough, irregular masses up to 4 feet in length. The zoaria of the western shelf area are also irregular, but usually are no more than 5 inches in length. It is probable that the zoaria of both groups formerly encrusted clumps of *Thalassia*."

Anastasia Formation, Okeechobee Group. Elias Sellards, in his early pioneer studies, was the first to recognize and name "the extensive deposit of coquina rock found along the east coast." With the best outcrops of this coquina rock being found on Anastasia Island, opposite St. Augustine, Sellards named this unit the Anastasia Formation. Around the same time, Sanford (1913) named an indurated sandy calcarenite from the Palm Beach County coastline, calling it the "Palm Beach Limestone". Later, Cooke and Mossom (1929) considered this sandier southern component to be simply a local facies and included Sanford's Palm Beach Limestone in Sellard's Anastasia Formation. In this book, we concur with Cooke and Mossom and recognize a single narrow formation, running down the coast from St. John's County to northernmost Broward County. The eastern coast of the late Tarantian Floridian Peninsula, from modern-day St. Augustine south to northern Broward County, was lined by long strands of beaches and transverse dunes. The wave-crushed shells and quartz sand of these beach deposits were cemented in subaerial conditions during subsequent sea level lows, forming the coquina beach rock of the Anastasia Formation.

Lithologic Description and Areal Extent. Throughout its range, the Anastasia Formation is relatively homogeneous, being composed of rounded molluscan shell fragments and quartz sand cemented together by calcite spar and iron oxide. Typically, the Anastasia consists of one or more beds of coquina, with shell fragments varying from large (2–5 cm) to small (0.5–1 cm). Often, large beach-worn gastropods are embedded in the matrix of small rounded shell fragments and flakes. Quartz sand amounts are variable, with some facies (such as at the stratotype) being composed almost entirely of cemented large fragments and unbroken shells and with other facies being almost a quartz sandstone, with the shell fragments reduced to small rounded grains. Along Palm Beach County, the quartz sand fraction is high, producing a dense sandy calcarenite with lesser amounts of finely crushed shell fragments. This facies of the Anastasia weathers quickly, being covered with microkarstic solution holes. Along Jupiter Island, Palm Beach County (particularly Jupiter Blowing Rocks, Figure 13.5),

Figure 13.3 View of an exposure of the Miami Formation in Coral Gables, Miami, Dade County, Florida, showing the cross-bedding seen in the oölitic shallow marine "dunes". Photo by Anton Oleinik.

the Anastasia forms a 3 m-high cliff riddled with wave "blow holes." Unweathered sections of the Anastasia are variable in color, with shell fragments ranging from golden orange to tan, to light grayish-brown, to gray. Often the shell fragments retain their original color, adding flecks of red, white, or black to the base color of the coquina. The Anastasia crops out all along the eastern coast of Florida, but rarely more than 5 km west of the Intracoastal Waterway (Cooke, 1945: 266) (Figure 13.1). The formation extends from St. John's County to northernmost Broward County. At some localities, such as at Bathtub Beach near Stuart, Martin County and at Jupiter Blowing Rocks, Palm Beach County, the Anastasia forms low cliffs and extensive raised beach rock terraces and pavements. South of this area to Hillsboro Inlet, the formation crops out as sporadic, small, low cliff-like features (such as at Gulf Stream and Red Reef Parks, Palm Beach County). Throughout its range, the Anastasia varies in thickness from 10 m in St. John's County, to 1 m along Flagler County, to 3–4 m at Jupiter Blowing Rocks, Palm Beach County, to over 7 m at the Palm Beach road cut, Palm Beach Island, Palm Beach County. The Anastasia Formation underlies Singer Island, Palm Beach Island, and the islands of Lake Worth in Palm Beach County.

Stratotype. The type locality is the stretch of coquina outcrops near the lighthouse on Anastasia Island, St. John's County, Florida.

Age and Correlative Units. Based on its interfingering with the uranium-dated Miami Formation and Coffee Mill Hammock Member of the Fort Thompson Formation, the Anastasia Formation is now known to be of mid-Tarantian Age. The beach sands and shell fragments were deposited at this time, but were later cemented in subaerial conditions during the late Tarantian sea level drop. The Anastasia is correlative with the Socastee Formation of the Carolinas, the Key Largo Formation of southern Florida, and the interfingering units mentioned previously.

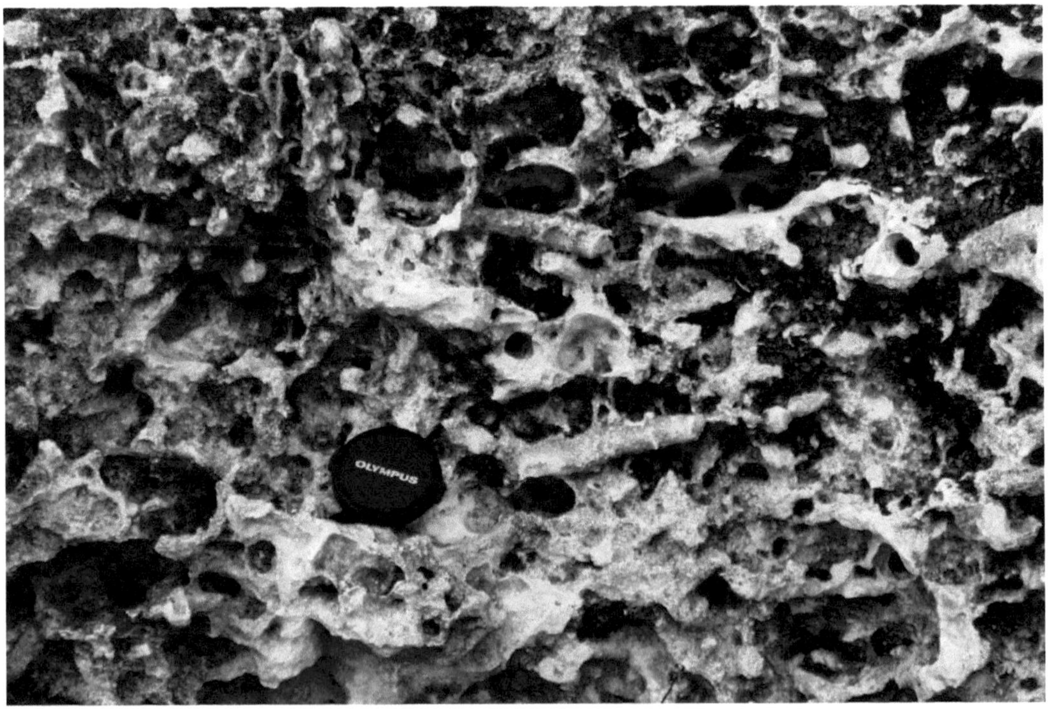

Figure 13.4 Close-up of an exposure of the Miami Formation in Miami, Dade County, Florida, showing interwoven branches of the cheilosome ectoproct *Schizoporella floridana*. Photo by Anton Oleinik.

Figure 13.5 View of a wave-eroded notch within the exposed Anastasia Formation at Jupiter Blowing Rocks Park, Jupiter, Palm Beach County, Florida. At high tide, these notches are submerged and, during wavy conditions, regularly produce powerful geyser-like jets of water. Photo by Anton Oleinik.

Anastasia Index Fossils. Because the formation is composed primarily of broken, wave-rounded shell fragments, few identifiable index fossils are encountered. At localities along Indian River, St. Lucie, and Martin Counties, large, fairly intact specimens of the busyconid gastropod *Sinistrofulgur sinistrum* have been found embedded in the shell fragment matrix. Rare fossilized specimens of the Ghost Crab, *Oxypode quadrata*, have been collected in Indian River County.

Holocene Formations. Three other minor sedimentary units were deposited within the Everglades area during the Holocene (past 11,700 years), but since these are not of marine origin, they will not be included in this book. These include:

1. **Lake Flirt Marl**, a heavily-indurated micritic limestone or partially-indurated layer of fine-grained calcarenite that is directly subjacent to the modern Everglades and was deposited in small freshwater lakes just prior to the formation of the Everglades (this unit also includes a "Flamingo Marl" facies that is found in the southern part of Everglades National Park).
2. **Pamlico Sand**, loose aeolianites that form the surface layer of quartz sand found all over eastern and southern Florida, including the large transverse dunes of coastal eastern Florida and adjacent inland areas.
3. **Everglades Peat**, a thick soil (up to 5 m) composed of cellulose fibers and plant material, stained black or dark brown, which formed within the paludal and palustrine environments of the Everglades wetlands prairies. Most of the plant matter derives from decomposed Saw Grass.

With the exception of the limited surficial outcrops of the Key Largo, Miami, and Anastasia formations, these three Holocene units completely cover all of the underlying geological units and obscure their lithologies and areal distributions. For details on the terrestrial Holocene deposits of the Everglades area, see Petuch and Roberts (2007: 193-197).

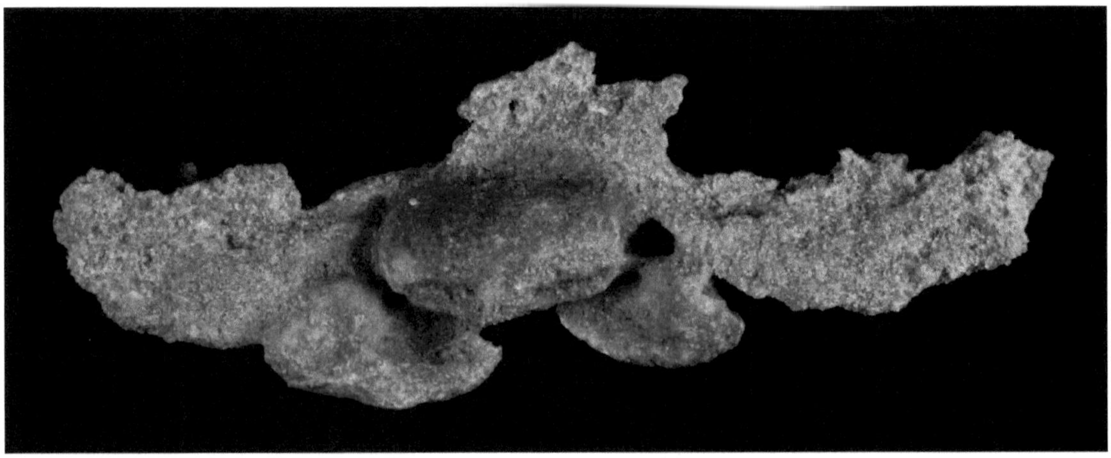

Oxypode quadrata (Fabricius, 1787) dorsal view.

REFERENCES

Applegate, A.V., April 28, 1986, Corkscrew field boosts South Florida drilling hopes, *Oil and Gas Journal*, 104–106.

Bottomley, R., Grieve, R.A.F., York, D., and Masaitis, V., 1997, The age of the Popigai impact event and its relationship to events at the Eocene-Oligocene boundary, *Nature* 388: 365–368.

Brooks, H.K., 1968, The Plio-Pleistocene of Florida, with special reference to the strata outcropping on the Caloosahatchee River, in Perkins, R.D., Ed., Late Cenozoic Stratigraphy of Southern Florida — A Reappraisal, with Notes on the Sunoco-Felda and Sunniland Oil Fields, pp. 3–42, *Guidebook, Second Annual Field Trip*, February 1968, Miami Geological Society, Coral Gables, FL.

Brooks, H.K., 1974, Lake Okeechobee, *in* Gleason, P.J., Ed., Environments of South Florida: Present and Past, *Miami Geological Society Memoir 2*, Miami Geological Society, Coral Gables, FL, pp. 256–286.

Campbell, L.D., 1993, *Pliocene Molluscs from the Yorktown and Chowan River Formations of Virginia*, Virginia Division of Mineral Resources, Publication 127, Charlottesville, VA, 259 pp.

Cooke, C.W., 1945, *Geology of Florida*, Geological Bulletin 29, Florida Geological Survey, State of Florida Department of Conservation, Tallahassee, FL, 339 pp.

Cooke, C.W. and Mansfield, W.C., 1936, Suwannee Limestone of Florida (abstract), *Geological Society of America Proceedings for 1935*, pp. 71–72.

Cooke, C.W. and Mossom, S., 1929, *Geology of Florida*, Florida Geological Survey Annual Report 20, Florida Geological Survey, Tallahassee, FL, pp. 29–227.

Cunningham, K.J., Locker, S.D., Hine, A.C., Bukry, D., Barron, J.A., and Guertin, L.A., 2003, Interplay of Late Cenozoic Siliciclastic and Carbonate response on the Southeast Florida Platform, *Journal of Sedimentary Research*, 73(1): 31–46.

Dall, W.H., 1887, Notes on the Geology of Florida, *American Journal of Science* (Series 3) 34: 161–170.

Dall, W.H., 1890, Contributions to the Tertiary Fauna of Florida, with especial reference to the Miocene Silex Beds of Tampa and the Pliocene Beds of the Caloosahatchie River. Part 1, *Transactions of the Wagner Free Institute of Science of Philadelphia* 3(1): 1–200.

Dall, W.H., 1892, Contributions to the Tertiary Fauna of Florida, with especial reference to the Miocene Silex Beds of Tampa and the Pliocene Beds of the Caloosahatchie River. Part 2, *Transactions of the Wagner Free Institute of Science of Philadelphia* 3(2): 201–474.

Dall, W.H., 1903, Contributions to the Tertiary Fauna of Florida, with especial reference to the Miocene Silex Beds of Tampa and the Pliocene Beds of the Caloosahatchie River. Part 6, Concluding the Work, *Transactions of the Wagner Free Institute of Science of Philadelphia* 3(6): 1219–1654.

Dall, W.H., 1915, A monograph on the molluscan fauna of the *Orthaulax pugnax* Zone of the Oligocene of Tampa, Florida, *U.S. National Museum Bulletin* 90: 1–173.

Douglas, M.S., 1947, *The Everglades: River of Grass* (revised ed.; 1992), Mockingbird Books, Marietta, GA, 308 pp.

Dowsett, H.J. and Cronin, T.M., 1990, High eustatic sea level during the middle Pliocene: evidence from the southeastern U.S. Atlantic Coastal Plain, *Geology* 18(5): 435–438.

DuBar, J.R., 1958, *Stratigraphy and Paleontology of the Late Neogene Strata of the Caloosahatchee River Area of Southern Florida*, Geological Bulletin 40, Florida Geological Survey, Florida State Board of Conservation, Tallahassee, FL, 267 pp.

DuBar, J.R., 1958a, Neogene stratigraphy of southwestern Florida, *Transactions of the Gulf Coast Association of Geological Societies* 8: 129–155.

DuBar, J.R., 1974, *Summary of the Neogene Statigraphy of southern Florida*, *in* Oaks, R.Q. and DuBar, J.R., Eds., Post-Miocene Stratigraphy, Central and Southern Atlantic Coastal Plain, Utah State University Press, Provo, UT, pp. 206–231.

Enos, P. and Perkins, R.D., 1977, *Quaternary Sedimentation in South Florida*, Geological Society of America Memoir 147, 198 pp.

Ganapathy, R., 1982, Evidence for a Major Meteorite Impact on the Earth 34 Million Years Ago: Implication for Eocene Extinctions, *Science* 216, pp. 885–886.

Glass, B.P. and Zwart, M.J., 1979, North American microtektites in deep sea drilling project cores from the Caribbean Sea and Gulf of Mexico, *Geological Society of America Bulletin* (Part 1), 90: 595–602.

Gleason, P.J., Ed., 1984, *Environments of South Florida: Present and Past II*, Miami Geological Society, Coral Gables, FL, 551 pp.

Haq, B.U., Hardenbol, J., and Vail, P.R., 1988, Mesozoic and Cenozoic chronostratigraphy and cycles of sea-level change, in Wilgus, C.K., Ed., Sea-level Changes — An Integrated Approach, *Society of Economic Paleontologists and Mineralogists Special Report* 42: 71–108.

Heilprin, A., 1886, Explorations on the West Coast of Florida and in the Okeechobee Wilderness, *Transactions of the Wagner Free Institute of Science of Philadelphia* 1: 366–506.

Hoerle, S.E., 1970, Mollusca of the "Glades Unit" of southern Florida. Part 2: List of the Molluscan Species from the Belle Glade Rock Pit, Palm Beach County, FL, *Tulane Studies in Geology and Paleontology* 8(1,2): 56–68.

Hoffmeister, J.E. and Multer, H.G., 1968, Geology and origin of the Florida Keys, *Geological Society of America Bulletin* 79: 1487–1502.

Hoffmeister, J.E., Stockman, K.W., and Multer, H.G., 1967, Miami limestone of Florida and its Recent Bahamian counterpart, *Geological Society of America Bulletin* 78: 175–190.

Hollister, S.C., 1971, New *Vasum* species of the subgenus *Hystrivasum*, *Bulletins of American Paleontology* 58(262): 209–304.

Hunter, M.E., 1968, Molluscan Guide Fossils in Late Miocene Sediments of Southern Florida, *Transactions of the Gulf Coast Association of Geological Societies* 18: 439–450.

Hunter, M.E., 1978, *What is the Caloosahatchee Marl? in* Hydrogeology of south-central Florida, Southeastern Geological Society Publication 20, pp. 61–88.

Johnson, L.C., 1888, The structure of Florida, *American Journal of Science* (Series 3), 36: 230–236.

Jones, D.S., MacFadden, B.J., Webb, S.D., Mueller, P.A., Hodell, D.A., and Cronin, T.M., 1991, Integrated geochronology of a classic pliocene fossil site in Florida: linking marine and terrestrial biochronologies, *The Journal of Geology* 99(5): 637–648.

Ketcher, K.M., 1992, Stratigraphy and Environment of Bed 11 of the "Pinecrest" Beds at Sarasota, Florida, in Scott, T.M. and Allmon, W.D., Eds., *Plio-Pleistocene Stratigraphy and Paleontology of Southern Florida*, Special Publication 36, Florida Geological Survey, Florida Department of Natural Resources, Tallahassee, FL, pp. 167–176.

King, K.C. and Wright, R., 1979, Revision of the Tampa Formation, west central Florida, *Transactions of the Gulf Coast Association of Geological Societies* 29: 257–262.

Klitgord, K.D., Popenoe, P., and Schouten, H., 1984, Florida: a Jurassic transform plate boundary, *Journal of Geophysical Research* 89: 7753–7772.

Krantz, D.E., 1991, A chronology of Pliocene sea level fluctuations, U.S. Atlantic Coastal Plain, *Quaternary Science Reviews* 10: 163–174.

Lyons, W.G., 1991, Post-Miocene species of Latirus Montfort, 1810 (Mollusca: Fasciolariidae) of southern Florida, with a review of regional marine biostratigraphy, *Bulletin of the Florida Museum of Natural History* 35(3): 131–208.

Lyons, W.G., 1992, Caloosahatchee-age and younger molluscan assemblages at APAC mine, Sarasota County, Florida, in Scott, T.M. and Allmon, W.D., Eds., *The Plio-Pleistocene Stratigraphy and Paleontology of Southern Florida*, Special Publication 36, Florida Geological Survey, Florida Department of Natural Resources, Tallahassee, FL, pp. 133–159.

MacFadden, B.J., 1995, Magnetic polarity stratigraphy and correlation of the Leisey Shell Pits, Hillsborough County, Florida, in Hulbert, R.C., Morgan, G.S., and Webb, S.D., Eds., *Paleontology and Geology of the Leisey Shell Pits, Early Pleistocene of Florida, Part 1*, Bulletin of the Florida Museum of Natural History, 37, Part 1(1–10): 107–116.

Mansfield, W.C., 1930, *Miocene Gastropods and Scaphopods of the Choctawhatchee Formation of Florida*, Geological Bulletin 3, Florida Geological Survey, Florida Department of Conservation, Tallahassee, FL, 142 pp.

Mansfield, W.C., 1931, Some Tertiary mollusks from southern Florida, *Proceedings of the U.S. National Museum* 79(21): 1–12.

Mansfield, W.C., 1932, *Pliocene fossils from limestone in southern Florida*, U.S. Geological Survey Professional Paper, 170: 43–56.

Mansfield, W.C., 1937, *Mollusks of the Tampa and Suwannee Limestones of Florida*, Bulletin 15, Florida Geological Survey, Florida Department of Conservation, Tallahassee, FL, 334 pp.

Mansfield, W.C., 1939, *Notes on the Upper Tertiary and Pleistocene Mollusks of Peninsular Florida*, Bulletin 18, Florida Geological Survey, Department of Conservation, Tallahassee, FL, 75 pp.

Matson, G.C. and Clapp, F.G., 1909, *A Preliminary Report on the Geology of Florida, with Special Reference to the Stratigraphy*, Florida Geological Survey Annual Report 2, Florida Geological Survey, Tallahassee, FL, pp. 25–173.

Maurasse, F. and Glass, B., 1976, Radiolarian stratigraphy and North American microtektites in Caribbean RC8–58: implications concerning Late Eocene radiolarian chronology and the age of the Eocene-Oligocene boundary, in Causse, R., Ed., *Transactions of the 7th Caribbean Geological Conference, Guadeloupe*, pp. 205–212.

McCaslin, J.C., April 28, 1986, NW Florida produces most of the state's oil, *Oil and Gas Journal* 103–104.

McGinty, T.L., 1940, New land and marine Tertiary shells from southern Florida, *The Nautilus* 53(3): 81–84.

McGinty, T.L., 1970, Mollusca of the "Glades" Unit of southern Florida: Part 1. Introduction and Observations, *Tulane Studies in Geology and Paleontology* 8(2): 53–56.

Meeder, J.F., 1979, *A Field Guide with Road Log to the Pliocene Fossil Reef of Southwest Florida*, Field Trip Guide (1979), Miami Geological Society, Coral Gables, FL, 19 pp.

Meeder, J.F., 1980, New information on Pliocene reef limestones and associated facies in Collier and Lee counties, Florida, in Gleason, P., Ed., *Water, Oil, and the Geology of Collier and Lee Counties*, Miami Geological Society, Coral Gables, FL, pp. 27–30.

Merritt, M.I., Myer, F.W., Sonntag, W.H., and Fitzpatrick, D.J., 1983, *Subsurface Storage of Freshwater in South Florida: A Prospectus*, U.S. Geological Survey Water Resources Investigations Report 83–4214, pp. 29–40.

Miller, W.L., 1987, *Lithology and the Base of the Surficial Aquifer System, Palm Beach County,* Florida, U.S. Geological Survey Water Resources Investigations Report 86–4067, 1 map.

Missimer, T.M., 1984, The geology of South Florida: a summary, *in* Gleason, P.J., Ed., *Environments of South Florida, Present and Past 2*, Miami Geological Society Memoir 2, Miami Geological Society, Coral Gables, FL, pp. 385–404.

Missimer, T.M., 1992, Stratigraphic relationships of sediment facies in the Tamiami Formation of southwest Florida: proposed intraformational correlations, in Scott, T.M. and Allmon, W.D., Eds., *The Plio-Pleistocene Stratigraphy and Paleontology of Southern Florida*. Special Publication 36, Florida Geological Survey, Florida Department of Natural Resources, Tallahassee, FL, pp. 63–92.

Morgan, G.S. and Hulbert, R.C., 1995, Overview of the geology and vertebrate biochronology of the Leisey Shell Pit Local Fauna, Hillsborough County, Florida, *in* Hulbert, R.C., Morgan, G.S., and Webb, S.D., Eds., *Paleontology and Geology of the Leisey Shell Pits, Early Pleistocene of Florida, Part 1*, Bulletin of the Florida Museum of Natural History, 37, Part 1(1–10): 1–92.

Mossom, D.S., 1925, *A Preliminary Report on the Limestones and Marls of Florida*, Florida Geological Survey Annual Report 16, Florida Geological Survey, Tallahassee, FL, pp. 27–203.

Olsson, A.A., 1967, *Some Tertiary Mollusks from South Florida and the Caribbean*, Paleontological Research Institute, Ithaca, NY, 61 pp.

Olsson, A.A. and Harbison, A., 1953, *Pliocene Mollusca of Southern Florida, with special reference to those from North Saint Petersburg*, Monographs of the Academy of Natural Sciences of Philadelphia, 8: 1–361.

Olsson, A.A. and Petit, R.E., 1964, Some Neogene Mollusca from Florida and the Carolinas, *Bulletins of American Paleontology* 47(217): 509–575.

Olsson, A.A. and Petit, R.E., 1968, Notes on Siphocypraea, *Bulletins of American Paleontology* 54(242): 279–289.

Osmond, J.K., Carpenter, J.R., and Windom, H.L., 1965, Th 230/U 234 age of the Pleistocene corals and oölites of Florida, *Journal of Geophysical Research* 70: 1843–1847.

Parker, G.G. and Cooke, C.W., 1944, *Late Cenozoic Geology of Southern Florida, with a Discussion of the Ground Water*, Florida Geological Survey Bulletin 27, Florida Geologic Survey, Florida Department of Conservation, Tallahassee, FL, 119 pp.

Parker, G.G., Ferguson, G.E., and Love, S.K., 1955, *Water Resources of Southeastern Florida*, U.S. Geological Survey Water-Supply Paper 1255, 965 pp.

Parodiz, J.J., 1988, A new species of *Siphocypraea* (Gastropoda:Cypraeidae) from the Neogene of southwest Florida, *Annals of the Carnegie Museum* 57(3): 91–97.

Parodiz, J.J. and Tripp, J.J., 1992, The Neogene Cassidae of southern Florida, with description of a new species of *Cassis* (Gastropoda:Prosobranchia), *Annals of the Carnegie Museum* 61(4): 317–325.

Petuch, E.J., 1976, An unusual molluscan assemblage from Venezuela, *The Veliger* 18: 320–325.

Petuch, E.J., 1979, A new species of *Siphocypraea* from northern South America, with notes on the genus in the Caribbean, *Bulletin of Marine Science* 29(2): 216–225.

Petuch, E.J., 1982, Notes on the Molluscan Paleontology of the Pinecrest Beds at Sarasota, Florida, with the description of *Pyruella*, a stratigraphically important new genus. *Proceedings of the Academy of Natural Sciences of Philadelphia* 134: 12–30.

Petuch, E.J., 1986, The Pliocene reefs of Miami: their geomorphological significance in the evolution of the Atlantic coastal ridge, southeastern Florida, U.S., *Journal of Coastal Research* 2(4): 391–408.

Petuch, E.J., 1987a, The Florida Everglades: a buried pseudoatoll? *Journal of Coastal Research* 3(2): 189–200.

Petuch, E.J., 1987b, A new *Ecphora* fauna from southern Florida, *The Nautilus* 10(4): 200–206.

Petuch, E.J., 1988a, *Neogene History of Tropical American Mollusks: Biogeography and Evolutionary Patterns of Tropical Western Atlantic Mollusca*, The Coastal Education and Research Foundation, Charlottesville, VA, 217 pp.

Petuch, E.J., 1988b, A new giant *Muricanthus* from the Pliocene of southern Florida, *Bulletin of Paleomalacology* 1(1): 17–21. (copies available at the Smithsonian Institution)

Petuch, E.J., 1989a, *Field Guide to the Ecphoras*, The Coastal Education and Research Foundation, Charlottesville, VA, 140 pp.

Petuch, E.J., 1989b, New species of *Malea* from the Pleistocene of southern Florida, *The Nautilus* 103(3): 92–95.

Petuch, E.J., 1990, New gastropods from the Bermont formation (middle Pleistocene) of the Everglades Basin, *The Nautilus* 104(3): 96–104.

Petuch, E.J., 1991, *New Gastropods from the Plio-Pleistocene of Southwestern Florida and the Everglades Basin*, Special Publication 1, W.H. Dall Paleontological Research Center, Florida Atlantic University, 85 pp. (copies available at the Smithsonian Institution and the Carnegie Museum).

Petuch, E.J., 1992, *Edge of the Fossil Sea: Life Along the Shores of Prehistoric Florida.* Bailey-Matthews Shell Museum, Sanibel, Florida. 80 pp.

Petuch, E.J., 1993, Patterns of Gastropod Extinction in the Plio-Pleistocene Okeechobean Sea of Southern Florida, in Zullo, V. *et al.*, Eds., *The Neogene of Florida and Adjacent Regions*, Proceedings of the Third Bald Head Island Conference on Coastal Plains Geology, Special Publication 37, Florida Geological Survey, Tallahassee, FL, pp. 73–85.

Petuch, E.J., 1994, *Atlas of Florida Fossil Shells (Pliocene and Pleistocene Marine Gastropods)*, The Graves Museum of Archaeology and Natural History, Dania, FL, and Spectrum Press, Chicago, IL, 394 pp.

Petuch, E.J., October 13, 1995, Molluscan diversity in the Late Neogene of Florida: evidence for a two-staged mass extinction, *Science* 270, pp. 275–277.

Petuch, E.J., 1996, *Calusacypraea*, a new, possibly Neotenic, genus of cowries from the Pliocene of southern Florida, *The Nautilus* 110(1): 17–21.

Petuch, E.J., 1997a, *Coastal Paleoceanography of Eastern North America.* Kendall/Hunt Publishing Company, Dubuque, Iowa. 373 pp.

Petuch, E.J., 1997b, A new gastropod fauna from an Oligocene back-reef lagoonal environment in west central Florida, *The Nautilus* 110(4): 122–138.

Petuch, E.J., 1998, New cowries from the Myakka Lagoon System and the Everglades Pseudoatoll (Pliocene Okeechobean Sea). *La Conchiglia* 30(288): 27–37.

Petuch, E.J., 2004, *Cenozoic Seas: The View from Eastern North America*, CRC Press, Boca Raton, FL, 308 pp.

Petuch, E.J., 2008, *The Geology of the Florida Keys and Everglades: An Illustrated Field Guide to Florida's Hidden Beauty.* Thomson Publishers, Mason, Ohio. 84 pp.

Petuch, E.J. 2012. Miocene Asteroid Impacts: Proposed Effects on the Biogeography and Extinction Patterns of Eastern North American Gastropods. in *Earth and Life: Global Biodiversity, Extinction Intervals, and Biogeographic Perturbations Through Time.* J.A. Talent, Editor. Springer Publishers, Dordrecht, London, New York. pp. 967-981.

Petuch, E.J, Berschauer, D.P., and Myers, R.F., 2018, *Jewels of the Everglades: The Fossil Cowries of Southern Florida.* Special Publication of the San Diego Shell Club, Encinitas, California. 248 pp.

Petuch, E.J. and Drolshagen, M., 2011, *Compendium of Florida Fossil Shells. Volume 1.* MdM Publishers, Wellington, Florida. 412 pp.

Petuch, E.J., Drolshagen, M., and Herndl, G., 2015, *Cone Shells of the Okeechobean Sea (Pliocene, Pleistocene)*. ConchBooks, Harxheim, Germany. 180 pp.

Petuch, E.J., Myers, R.F., and Berschauer, D.P., 2015, *The Living and Fossil Busycon Whelks: Iconic Mollusks of Eastern North America.* Special Publication of the San Diego Shell Club, Encinitas, California. 194 pp.

Petuch, E.J. and Roberts, C.E., 2007. *The Geology of the Everglades and Adjacent Areas.* CRC Press, London, Boca Raton. 212 pp.

Pilsbry, H.A., 1934, Review of the Planorbidae of Florida, with notes on other members of the family, *Proceedings of the Academy of Natural Sciences of Philadelphia* 86: 29–66.

Poag, C.W., 1999, *Chesapeake Invader: Discovering America's Giant Meteor Crater*, Princeton University Press, NJ, 183 pp.

Poag, C.W., Koeberl, C., and Reimold, W.U., 2004, *The Chesapeake Bay Crater: Geology and Geophysics of a Late Eocene Submarine Impact Structure*, Springer-Verlag, New York, 522 pp.

Portell, R.W., Schindler, K.S., and Nicol, D., 1995, Biostratigraphy and Paleoecology of the Pleistocene Invertebrates from the Leisey Shell Pits, Hillsborough County, Florida, in Hulbert, R.C., Morgan, G.S., and Webb, S.D., Eds., *Paleontology and Geology of the Leisey Shell Pits, Early Pleistocene of Florida, Part 1*, Bulletin of the Florida Museum of Natural History, 37, Part 1(1–10): 127–149.

Powars, D.S., 2000, *The effects of the Chesapeake Bay impact crater on the geologic framework and the correlation of hydrogeologic units of Southeastern Virginia, south of the James River*, U.S. Geological Survey Professional Paper 1622, U.S. Geological Survey, 53 pp.

Powars, D.S. and Bruce, T.S., 1999, *The effects of the Chesapeake Bay impact crater on the geological framework and correlation of hydrogeologic units of the Lower York-James Peninsula*, VA, U.S. Geological Survey Professional Paper 1612, U.S. Geological Survey, 82 pp.

Puri, H.S. and Vernon, R.O., 1964, *Summary of the Geology of Florida*, Special Publication 5, Florida Geological Survey, Florida Department of Natural Resources, Tallahassee, FL, 312 pp.

Puri, H.S. and Winston, G.O., 1974, *Geologic Framework of the High Transmissivity Zones in South Florida*, Special Publication 20, Bureau of Geology, Florida Department of Natural Resources, Tallahassee, FL, pp. 41–58, cross section C-D.

Randazzo, A.F. and Jones, D.S., Eds., 1997, *The Geology of Florida*, University Press of Florida, Gainesville, FL, 327 pp.

Sanfilipo, A.L., Riedel, W.R., Glass, B.P., and Kyte, F.T., 1985, Late Eocene microtectites and radiolarian extinctions on Barbados, *Nature* 314: 613–615.

Sanford, S., 1909, *The Topography and Geology of Southern Florida*, Florida Geological Survey Annual Report 2, Florida Geological Survey, Tallahassee, FL, pp. 175–231.

Sanford, S., 1913, *Geology and Ground Waters of Florida: Southern Florida*, U.S. Geological Survey Water-Supply Paper 319, pp. 42–64.

Scott, T.M., 1988, *The Lithostratigraphy of the Hawthorn Group (Miocene) of Florida*, Bulletin 59, Florida Geological Survey, Florida Department of Natural Resources, Tallahassee, FL, 148 pp.

Scott, T.M., 1992, Coastal Plains Stratigraphy: the dichotomy of biostratigraphy and lithostratigraphy — a philosophical approach to an old problem, in Scott, T.M. and Allmon, W.D., Eds., *The Plio-Pleistocene Stratigraphy and Paleontology of Southern Florida*, Special Publication 36, Florida Geological Survey, Florida Department of Natural Resources, Tallahassee, FL, pp. 21–25.

Scott, T.M., 1997, Miocene to holocene history of Florida, in Randazzo, A.F. and Jones, D.S., Eds., *The Geology of Florida*, University Press of Florida, Gainesville, FL, pp. 57–67.

Sellards, E.H., 1919, *Geologic Sections across the Everglades of Florida*, Florida Geological Survey Annual Report 12, Florida Geological Survey, Tallahassee, FL, pp. 105–141.

Smith, M., 1936, New tertiary shells from Florida, *Nautilus* 49(4): 135–139.

Stubbs, S.A., 1940, Pliocene mollusks from a well at Sanford, Florida, *Journal of Paleontology* 14(5): 510–514.

Swayze, L.J. and Miller, W.L., 1984, *Hydrogeology of a Zone of Secondary Permeability in the Surficial Aquifer of Eastern Palm Beach County, Florida*, U.S. Geological Survey Water-Resources Investigations Report 83-4249 (in cooperation with Palm Beach County), U.S. Geological Survey, 38 pp.

Tucker, H.I. and Wilson, D., 1932, Some new or otherwise interesting fossils from the Florida Tertiary, *Bulletin of American Paleontology* 18(65): 1–24.

Tucker, H.I. and Wilson, D., 1933, A second contribution to the Neogene paleontology of South Florida, *Bulletin of American Paleontology* 18(66): 1–21.

Vail, P.R. and Hardenbol, J., 1979, Sea-level changes during the Tertiary, *Oceanus* 23(3): 71–79.

Vail, P.R., Mitchum, R.M., and Thompson, S., 1977, Global cycles of sea level changes, *in* Payton, C.E., Ed., *Seismic Stratigraphy: Applications to Hydrocarbon Exploration*, American Association of Petroleum Geologists Memoir 26, pp. 83–97.

Vaughan, T.W., 1919, *Fossil corals from Central America, Cuba, and Porto Rico, with an account of the American Tertiary, Pleistocene, and Recent coral reefs*, U.S. National Museum Bulletin, 103: 189–524.

Vermeij, G.J. and Vokes, E.H., 1997, Cenozoic Muricidae of the western Atlantic Region. Part 12: The Subfamily Ocenebrinae (in part), *Tulane Studies in Geology and Paleontology* 29(3): 69–18.

Vokes, E.H., 1963, Cenozoic Muricidae of the western Atlantic Region. Part 1: *Murex sensu stricto, Tulane Studies in Geology* 1(3): 93–123.

Vokes, E.H., 1965, Cenozoic Muricidae of the western Atlantic Region. Part 2: *Chicoreus sensu stricto* and *Chicoreus (Siratus)*, *Tulane Studies in Geology* 3(4): 181–204.

Vokes, E.H., 1966, The genus *Vasum* (Mollusca: Gastropoda) in the New World, *Tulane Studies in Geology* 5(1): 1–36.

Vokes, E.H., 1975, Cenozoic Muricidae of the western Atlantic Region. Part 6: *Aspella* and *Dermomurex*, *Tulane Studies in Geology and Paleontology* 11(3): 121–162.

Vokes, E.H., 1976, Cenozoic Muricidae of the western Atlantic Region. Part 7: *Calotrophon* and *Attiliosa*, *Tulane Studies in Geology and Paleontology* 12(3): 101–136.

Waldrop, J.S. and Wilson, D., 1990, Late Cenozoic stratigraphy of the Sarasota area, *in* Allmon, W.D. and Scott, T.M., Eds., *Plio-Peistocene Stratigraphy of South Florida*, Southeastern Geological Society Annual Fieldtrip Guidebook 31, 221 pp.

Ward, L.W., 1992, Diagnostic Mollusks from the APAC Pit, Sarasota, Florida, in Scott, T.M. and Allmon, W.D., Eds., *Plio-Pleistocene Stratigraphy and Paleontology of Southern Florida*, Special Publication 36, Florida Geological Survey, Florida Department of Natural Resources, Tallahassee, FL, pp. 161–165.

Weisbord, N.E., 1974, Late Cenozoic Corals of South Florida, *Bulletins of American Paleontology* 66 (285): 259–544.

White, W.A., 1970, *The Geomorphology of the Florida Peninsula*, Florida Geological Survey Bulletin 51, Tallahassee, FL, pp. 1–185.

An artistic rendering of a coral bioherm that grew along the Palm Beach Archipelago of the Loxahatchee Subsea during the Calabrian Pleistocene. Here the grinning tun shell, *Malea petiti*, crawls on coral rubble along the left, and a large cowrie, *Macrocypraea (Lorenzicypraea) spengleri*, hides under a ledge of *Siderastrea* corals on the right. This assemblage is typical of the coral reef facies of the Holey Land Member of the Bermont Formation. Taken from Petuch, 1992.

INDEX

A

B

C

D

E

F

G

H

I

J

K

L

M

N

O

T

U

V

W

Z

The muricid gastropod, *Globecphora parvicostata* (Pilsbry, 1911), length 64 mm. This distinctive Pliocene gastropod occurs in the Rushmere Member of the Yorktown Formation of Virginia and North Carolina, the Goose Creek Formation of South Carolina, and, rarely, in the Kissimmee facies of the Pinecrest Member of the Tamiami Formation. The few known Florida specimens have only been collected during channel dredging of the Kissimmee River.

Close-up view of the giant cowrie, *Macrocypraea (Lorenzicypraea) spengleri*, embedded in the dense micritic limestone of the Holey Land Member of the Bermont Formation at the Griffin Brothers Pit, Holey Land Wildlife Management Area, southwestern Palm Beach County.

ABOUT THE AUTHORS

Edward J. Petuch was born in Bethesda, Maryland, in 1949. Raised in a Navy family, he spent many of his childhood years collecting living and fossil shells in such varied localities as Chesapeake Bay, California, Puerto Rico, and Wisconsin. His early interests in malacology and oceanography eventually led to BA and MS degrees in zoology from the University of Wisconsin-Milwaukee. During his MS thesis research, Petuch concentrated on the molluscan biogeography of West Africa, traveling extensively in the Canary Islands, Western Sahara, Senegal, Gambia, Sierra Leone, Ivory Coast, and Cameroons. During this time, he also conducted research on the molluscan ecology of both coasts of Mexico and the Great Barrier Reef of Belize.

Continuing his education, Petuch studied marine biogeography and malacology under Gilbert Voss and Donald Moore at the Rosenstiel School of Marine and Atmospheric Sciences at the University of Miami, where he received a full scholarship. During this time, his doctoral dissertation research involved intensive field work in Costa Rica, Colombia, Venezuela, Barbados, the Grenadines, and Brazil, where he often went to sea with the local shrimpers for weeks at a time. After receiving his PhD in oceanography in 1980, Petuch was invited to conduct two years of postdoctoral research, funded by the National Science Foundation, with Geerat Vermeij at the University of Maryland. While there, he also held a research associateship with the Department of Paleobiology at the National Museum of Natural History, Smithsonian Institution, under the sponsorship of Thomas Waller, and conducted field work in the Plio-Pleistocene fossil beds of Florida and North Carolina and the Miocene fossil beds of Maryland and Virginia.

Petuch has also collected and studied living mollusks in Australia, Papua-New Guinea, Fijis, French Polynesia, Japan, the Bahamas, Nicaragua, and Uruguay. This research has led to the publication of over 350 scientific papers and the discovery and description of over 1,500 new species of mollusks and over 100 new genera. His previous 21 books are well-known reference texts in the malacological and paleontological communities, and some of the better known include: *Tropical Marine Mollusks: An Illustrated Biogeographical Guide* (2021), *Jewels of the Everglades: The Fossil Cowries of Southern Florida* (2018), *The Living and Fossil Busycon Whelks: Iconic Mollusks of Eastern North America* (2015), *Cone Shells of the Okeechobean Sea* (2015), *Molluscan Communities of the Florida Keys and Adjacent Areas: Their Ecology and Biodiversity* (2014), *Biogeography and Biodiversity of Western Atlantic Mollusks* (2013), *Molluscan Paleontology of the Chesapeake Miocene* (2010), *The Geology of the Everglades and Adjacent Areas* (2007), and *Cenozoic Seas: The View from Eastern North America* (2004). Currently, Petuch is a

Professor Emeritus in the Department of Geosciences, Florida Atlantic University in Boca Raton, Florida where, for thirty years, he taught undergraduate classes in oceanography, paleontology, and physical geology, and graduate classes in paleoecology and paleoceanography. He currently resides in Jupiter, Florida, with his wife Linda, where they both enjoy visits from their three children and their families.

David P. Berschauer was born in Rockville Center, New York, in 1964, and spent his youth collecting shells in such varied localities as California, New York, Florida, Washington, and Mexico. His early interests in natural history, malacology, and marine biology eventually led to a BS in biology at the University of California-Irvine, an advanced marine invertebrate zoology course at Washington State University's Friday Harbor Marine Lab, and studies towards the pursuit of a graduate degree in marine biology at Florida State University in Tallahassee, Florida. While still an undergraduate, Berschauer performed field field biology research, published a number of research papers and gave scientific presentations at national conferences. He subsequently switched career paths and attended Southwestern University School of Law in Los Angeles, California, earning his Juris Doctorate in 1991. Although having developed a legal career, he has kept malacology as a lifetime avocation and has put together a sizeable research collection and personal museum of molluscan specimens.

© Morgan Taylor Photography, with permission

Over his entire professional life, Berschauer continued to pursue his passion for marine biology, and collecting and studying marine organisms. In his spare time, he has developed and published a relational database software program to aid in the organization and maintenance of a systematic collection. Although originally designed for malacology, the program is applicable to entomology and other aspects of systematic zoology. Now residing in Laguna Hills, California, Berschauer is an active member of the San Diego Shell Club and is the Editor-in-Chief of the journal, *The Festivus*. He is also well known for his natural history and shell photography, with a multitude of high-quality examples seen throughout this book. Besides being the author of many important scientific papers on molluscan systematics, Berschauer has described and named over 100 new species of gastropods and is also the co-author of several recent books on mollusks, including *The Living and Fossil Busycon Whelks: Iconic Mollusks of Eastern North America* (2015), *Jewels of the Everglades: The Fossil Cowries of Southern Florida* (2018), and the well-received *Sea Shells of Southern California: Marine Shells of the Californian Province* (2018), and *Tropical Marine Mollusks: An Illustrated Biogeographical Guide* (2021). Between court cases and an active law practice, Berschauer and his wife Felicia find time to enjoy traveling on cruise ships and exploring.